Elements of
YACHT DESIGN

AT CLOSE QUARTERS!

50-Footers *Mystic* and *Iroquois II*

Elements of
YACHT DESIGN

The Original Edition of the Classic Book on Yacht Design

NORMAN L. SKENE

Introduction by Maynard Bray

SHERIDAN HOUSE

Essex, Connecticut

SHERIDAN HOUSE

An imprint of Globe Pequot, the trade division of
The Rowman & Littlefield Publishing Group, Inc.
4501 Forbes Blvd., Ste. 200
Lanham, MD 20706
www.rowman.com

Distributed by NATIONAL BOOK NETWORK

British Library Cataloguing in Publication Information available

Library of Congress Cataloging-in-Publication Data

ISBN 978-1-4930-7601-7 (paper : alk. paper)
ISBN 978-1-4930-7602-4 (electronic)

♾™ The paper used in this publication meets the minimum requirements of American
National Standard for Information Sciences—Permanence of Paper for Printed Library
Materials, ANSI/NISO Z39.48-1992.

INTRODUCTION

W hen asked if I thought this book any longer had value, I answered with a resounding "yes!" There are two reasons for my enthusiasm. The writing is succinct and based upon the same engineering principles and mathematical equations in use today. Furthermore, it has become a classic in its field, available (rarely) only through secondhand book dealers. It's been out of print for over 50 years.

Anyone taking up yacht designing these days relies upon one or another computer program to grind out the once-laborious calculations, those cut-and-dry stability calculations, especially, and this saves loads of time. But the formulas remain unchanged, and this book lists nearly all of them, along with a terse description of relationships between the factors that plug into them. Skene wasn't one to palaver long on any one theme, and he was not a particularly engaging writer. He created a reference book that's useful partly because of its brevity. If readers need more explanation (or more entertainment), there are other sources.

L. Francis Herreshoff's *Common Sense of Yacht Design*, for example, is mostly a non-technical discussion touching only lightly upon mathematics. It's a wonderful book, but of a completely different character from this one. Norman Skene and Francis Herreshoff knew each other well, both professionally and socially. Both resided on Massachusetts's North Shore. Skene's book had been in print at least 40 years (it first came out in 1904) before Herreshoff wrote his. I don't know for certain, but I suspect that Herreshoff, who considered yacht design more art than science, was motivated to write because of Skene's book—a book whose MIT-trained author believed yacht designers should be grounded in science and mathematics. This difference in philosophy resulted in two classic reference works that should be on every yacht designer's bookshelf.

Elements of Yacht Design may also have catalyzed Howard Chapelle to write *Yacht Designing and Planning* in 1936. That book's stated purpose was to show beginners and students the practical steps in designing on the drawing board. Chapelle's book is less readable than Herreshoff's, and less mathematical than Skene's. But there's some great material between its covers, written and illustrated by one of the world's most prolific marine draftsmen, whose drawings are nothing short of elegant.

Norman Locke Skene wrote *Elements of Yacht Design* the year after he graduated from college in 1901. First published in 1904, it was one of several books put out by The Rudder Publishing Company that began with the word "Elements," and sold initially for $2.00. Skene's little book grew in size as his career matured and his book gained acceptance. That career led him to work with W.

Starling Burgess, first with Burgess's airplane company and later with the yacht design firm of Burgess, Swasey & Paine. L. Francis Herreshoff worked there as well. Skene then worked briefly for Herreshoff who claimed that Norman Skene was about the fastest draftsman he ever knew. Later he became a partner in Paine, Belknap & Skene, designers of the America's Cup contender *Yankee* and other notable yachts, where he stayed until his untimely death in 1932. His *Elements* book, meanwhile, continued through many editions, curiously ending up with Kennedy Brothers, publishers of *The Rudder*'s rival magazine *Yachting*. Under that aegis, the book appears to have been reborn between hard covers to go through several more editions, the last two published posthumously. Its last and sixth edition came out in 1938 and was reprinted several times.

Over twenty years later, in 1962, when the "thoroughly revised and up-to-date" edition of the *Elements of Yacht Design* came out it was authored by Francis S. Kinney, and bore no resemblance to what Skene had written, even though it was entitled *Skene's Elements of Yacht Design*. By no means am I maligning what Kinney did; his is a wonderful treatise and I use it often. But it's hard to figure why he or publisher Dodd, Mead & Company chose to relate Kinney's book to Skene's earlier work. Kinney's is more than good enough to stand on its own. Because both Skene and Kinney approached yacht design and wrote about it differently, both of their books proved popular.

Kinney described how he designed the yachts that came out under his own name and described practices at the design offices of P.L. Rhodes and Sparkman & Stephens where he had worked. Included in *Elements* for the first time are two valuable scantling rules for wooden hulled yachts: the one developed by N.G. Herreshoff for the New York Yacht Club in 1927, and another by premier yacht builder Henry B. Nevins. Both rules are detailed and contain an exceptional store of information about establishing the size and proportions of the individual pieces making up a wooden yacht.

In a similar way, in this republished sixth edition of Skene's original *Elements of Yacht Design*, you'll find the chapter on racing contains the handicapping rules of the day. There were several. The Universal Rule created the letter boats like R-boats, S-boats, and J-boats. The International Rule spawned the meter boats, as in the 8, 10, and 12-meter sloops. The Scharen-Kreuzer Rule developed the square meter boats like the 22 and 30 square meters. And the Cruising Club of America (CCA) Rule resulted in the handsome ocean racing and cruising sloops and yawls of the 1930s, 40s, and 50s. Granted, that material is obsolete because those handicapping rules are no longer in effect for new designs. Nevertheless, it is of great historical interest. It explains why sailing yachts designed for racing were shaped and rigged the way they were, and the various ways in which yachts of varying size and type could compete on a level playing field.

On the powerboat front, Skene's original (and very well handled text) was supplemented for this sixth edition by George Crouch who was a leading designer of fast racing powerboats and at one time head of Webb Institute. He authored the chapter on planing hulls entitled Hydroplanes, and discussed the various speed-giving factors that make them go.

Some of Norman Skene's designs of catboats, launches, other watercraft, and some of his technical articles appeared in *The Rudder* or *Forest & Stream* magazines shortly after he graduated from MIT as a naval architect and marine engineer, and while he was designing under his own name. They stopped at about 1912, with a single exception. In the August 1923 issue of *The Rudder*, Skene wrote a piece entitled "Building Plans of Walrus, Eskimo Kayak." Walrus was a canvas-covered version of a genuine South Greenland kayak that he had measured, then made wider for more stability. Of kayaking, he wrote, "Clad in an abbreviated bathing suit and equipped with a light double-bladed paddle, one can travel long distances without fatigue in this easily-driven craft. An occasional plunge overboard adds to the zest of kayaking." How ironic that only a decade later, in June of 1932, Skene drowned in one of his beloved kayaks.

On his passing, *The Rudder* obituary called Skene one of America's foremost yacht designers. This book is his legacy.

As you read and study this book, you should keep in mind that not all of it remains valid, as there have been countless technological improvements over the past 60 years. New materials such as synthetic sailcloth and running rigging, carbon fiber masts, and fiberglass hulls alter design parameters and expand possibilities. For this reason, *Elements of Yacht Design*, because of its age, no longer contains everything a designer needs to know. But you'll find most of it here, and if supplemented by more recent books on the same subject, I think you'll find yourself browsing through it often as one of your primary references.

Maynard Bray
Brooklin, Maine
2001

PREFACE TO SIXTH EDITION

SINCE the publication of the fifth edition of this book, in 1935, a number of changes have been made in the measurement rules for racing yachts, both here and abroad. These changes have been incorporated in Chapter XII to bring it down to date. Chapter XVI, on the hydroplane, has been entirely rewritten by Mr. George F. Crouch, the eminent naval architect who has designed many of the fastest hydroplanes built in the last twenty-five years. Considerable material has been added to Chapter XVIII and minor corrections have been made throughout the book. It is hoped that the new edition will be as helpful to the naval architect, both professional and amateur, as have the earlier ones.

CHARLES H. HALL.

PREFACE TO FIFTH EDITION

YACHT designing is a most fascinating branch of engineering and one in which a very large number of people are interested. The rudiments of the art are not difficult to grasp and a study of the subject is of definite advantage to anyone interested in any phase of yachting.

This book is intended to be a practical and concise presentation of some of the operations involved in designing yachts of all types. Cumbersome and impractical methods which are so often found in more pretentious works on naval architecture have been avoided. Those presented have been in everyday use by the author. The thirty-footer design used for purposes of illustration is obsolete in type, having been designed many years ago, but serves the purpose of illustrating the various operations involved in designing a yacht.

Nearly every yacht nowadays, aside from racing yachts, is equipped with power. On this account the problems of speed, power and suitable propeller exist in nearly every design and special attention is given to these matters in this edition. The illustrations as a rule have been selected to emphasize some point in the text. The designer's name appears beneath the picture at the right and the water line length at the left. A great deal of original material has been introduced which it is hoped will make the book of definite use to naval architects of all degrees — from the boat owner who is interested only superficially in the art of design to the practicing professional.

NORMAN L. SKENE.

CONTENTS

Elements of
YACHT
DESIGN

ELEMENTS OF YACHT DESIGN

CHAPTER I

GENERAL DISCUSSION

THERE is a wonderful fascination about yacht designing because of the opportunity generally afforded the designer for experiment toward improvement of type, and for the expression of his genius practically unhampered by the many considerations which closely restrict the efforts of the designer of commercial vessels.

Another element of fascination is the complexity of the problem of providing the perfect yacht for a given set of conditions. The antagonistic natures of speed, seaworthiness, large cabin accommodations and beauty, with the varying and uncertain effects of waves, change of heel and trim, cut of sails, etc., takes the problem, especially in the case of the sailing yacht, out of the category of strictly engineering problems such as those of the aeroplane, locomotive or steamship, in which results can be predicted with great exactness by mathematics based on laboratory work.

REQUISITES OF SUCCESSFUL DESIGNER

It must not be inferred that science is not an important aid in designing any kind of a yacht, for it is first in the requisites for consistent success, but with it must be blended natural genius, imagination and much practical experience in handling and building boats. It is the combination of these qualifications that enables a designer to do good work. It is not enough to be fond of boats and full of inspiration — such is distinctly the amateur; it is not enough to have had years of experience at sea and in the boat shop — some ridiculous models are produced by old sailors who keenly appreciate a good boat but cannot produce one; it is not enough to know all about resistance, displacement, stability, etc. — for the purely scientific designer may blunder on many practical considerations.

The practical designer will often drift into some particular branch of yacht architecture as a specialty, such as small racers, large power cruisers, heavy cruising type schooners, high speed motor boats, etc., for with experience in one particular line comes proficiency, reputation and increased business in that specialty. Every boat is an experi-

3

ment, but in producing a special type the designer has a great advantage in stepping with confidence, born of experience, from one design to another with intelligent improvement in each succeeding boat.

This brings out the importance of studying thoroughly the performance of each boat and noting carefully any deficiencies for correction in the next boat. The designer who is engrossed only by the problem of the design and production of a yacht, and when it is done regards it

"GERTRUDE THEBAUD"

as a closed book and turns all his attention to producing other inspirational designs, can never be classed as a consistently practical yacht architect.

ORIGINALITY IMPERATIVE

Originality, based on one's own study and experimental work, is really the keynote of success. He who does things in a certain way because others are doing it and always have done it that way contributes little to the advancement of the art. One should study constantly the work of others, but rather with an eye to discovering what not to do than what to copy.

Yacht designing, with its comparative freedom from hampering restrictions, should lead the world of marine construction in development of form for speed, seaworthiness and gracefulness. This is the case with sailing craft — witness the great improvement in fishing vessels which now closely approach yacht form.

Improvements developed in yacht work should exert greater influence on commercial sailing vessels than has hitherto been the case. It is generally believed that sailing ships are practically a thing of the past and are doomed to extinction. I do not agree to this. The wind is an unfailing source of power, while the supply and cost of fuel for mechanically propelled vessels in the years to come are matters of great uncertainty. Even today I believe that for long voyages and

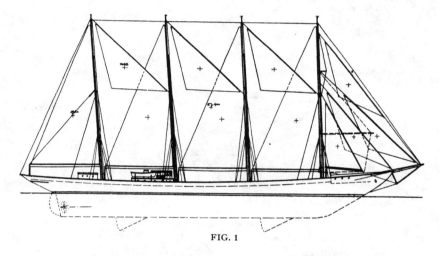

FIG. 1

bulky cargoes, an improved type of sailing vessel or auxiliary could compete profitably with motor ships.

Among the improvements which progressive naval architects may apply to sailing vessels are: Better speed to windward by means of radical improvements in form and rig and the use of centerboards or bilgeboards; economy in man power by free use of electric auxiliaries, and increased safety by utilization of automatic steerers, stabilizers and modern navigational aids such as are installed on many modern steam and motor ships. Where a portion of a voyage is necessarily through a region where head winds prevail, such as rounding Cape Horn from east to west, the matter of weatherliness is of utmost consequence.

The ocean sailing vessel of the future may possibly be of the modified fisherman form and rig, fitted with a Diesel driven electric generator plant which will supply current for the screw, windlass, pumps, winches, etc. All sheets and halliards may be of wire, each hauled by electric motor driving a drum on which all the slack is reeled. She may be equipped with gyroscopic automatic steering mechanism controlling a power rudder gear. Controls for all motors may be located at the central station where the officer of the deck can hoist, reef or trim sail and control the course, hoist anchor or centerboard, and start or stop the auxiliary screw, all from the central instrument board. In such a vessel a small crew would be required and the superior weatherliness, steadiness in steering and accuracy of navigating instruments will greatly increase the average speed and the safety, bringing the ship into competition with the power driven freighter.

Figure 1 is a suggestion for such a vessel. A study of long voyages of

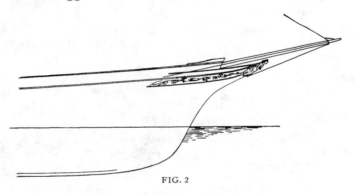

FIG. 2

sailing ships shows that the low average speed is due to calms or head winds over a comparatively small portion of the voyage. In the case of one voyage of a certain clipper ship, it was found that the fuel required to boost speed to nine knots on the light weather days was only one-eighth that which would have been required by a motor ship to make the same average speed, under power alone. The fuel saving figures more than twice the overhead on the rig. The added safety and ability to make a schedule put the auxiliary ship on 'a par with the motor ship.

In large power yachts designs have, curiously enough, always copied the commercial steamers of a decade or so previous. Until the advent of the Diesel yacht the large power yacht always had the clipper bow and short bowsprit, which prevailed on the best steamers of the middle of the 19th century (Fig. 2). Most Diesel and steam yachts were then built with the plumb stem and elliptical stern combination affected by commercial steamships following the clipper bow era. Progressive steamship designers have practically abandoned the elliptical knuckle stern in favor of the more buoyant and more easily constructed types of sterns shown in Figure 3. A yacht designer may be more justifiably influenced by the development of naval vessels, where sentiment has little part in the determination of hull form and proportions, these

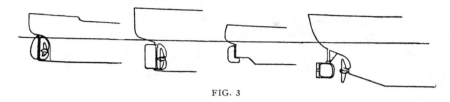

FIG. 3

being the product of extensive scientific investigation tempered by practical experience.

NAVAL DESIGNERS BOLDLY PROGRESSIVE

The battleship bow shown in Fig. 4 impresses one at first as a ruthless monstrosity, but on careful study it is found to be a wonderful improvement from the standpoint of efficiency over bows previously used both in above and below water form, the bulbous forefoot being a most remarkable innovation, one which could not have been devel-

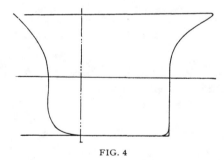

FIG. 4

oped without exhaustive model testing. This bow, in its above water form, could well be adapted to yacht work, for it is justified by improved speed and seaworthiness. Such a bow is employed in modified form in the sketch for a 115-foot water line power yacht (Fig. 5).

It is interesting to note a tendency on the part of the best designers to return to the clipper bow type on large power yachts. This form has intrinsic beauty and seems especially suited to the luxurious pleasure vessel.

DESIGNS ARE A COMPROMISE

There are four general characteristics sought after in yacht design: seaworthiness, large cabin accommodations, beauty and high speed. These properties do not readily combine; in fact, seldom do we see a successful combination of more than two of them. The best the designer can expect to do is to embody in his design the qualities especially desired, treating the other features in such a manner as to render their deficiencies as inconspicuous as possible.

The design of a yacht is too often regarded solely as an engineering problem. There are certain conditions to be met, such and such a speed must be made, certain stresses withstood, certain accommodations provided. The skilled designer produces a craft to fulfill these conditions following methods that are well understood. After working

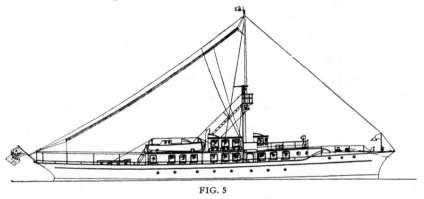

FIG. 5

into his design the above useful qualities he considers the matter of beauty of design, the purely architectural side of the problem, the elusive factor that crowns or damns the whole creation.

THE AESTHETIC ASPECT OF DESIGNING

What constitutes beauty in marine design? What rules must we study and follow? Unfortunately, this side of naval architecture has received scant consideration in standard works on the subject. Every designer is practically a law unto himself in such matters, and his knowledge of what makes for beauty is largely intuitive. There is no such wealth of tradition, no such definitely developed standards of style as exist in the architecture of buildings; nor is this strange, for a vessel must fulfill many most practical considerations and the purely aesthetic features must come last. There are various architectural features, standards of days gone by, such as quarter galleries, high poops, poop lanterns, gun ports, figure heads and the clipper bow which may be classed as belonging to the aesthetic phase of naval architecture but which, for practical reasons, are now all but obsolete.

Some of these features can occasionally be used for special requirements, such as a power houseboat in which a picturesque and distinctive appearance is sought.

UTILITY, SYMMETRY AND HARMONY

Although there are no definitely formulated rules for the attainment of beauty in design, I believe the fundamental principles governing the subject may be laid down somewhat as follows:

To begin with, the keynote of the matter is *utility*. Every feature of the design should be an expression of some useful purpose; otherwise, it is not justified. Any innovation which constitutes a distinct improvement, however strange and repellent it may seem at first, soon becomes acceptable from the standpoint of appearance. But if it is false, added merely for looks, away with it! It cannot continue to be considered good taste for it does not fulfill the condition of utility.

The design in its entirety should be a frank, vigorous declaration of the use to which the boat is to be put. If she is a cruiser, every line should tend to give the impression of strength, seaworthiness and comfort; if a racer, the refinement of design and construction should indicate speed; if a working boat, her sturdy lineaments should proclaim her commercial employment.

Another important principle is *symmetry*. It has extensive application in every phase of the design. Symmetry of distribution of displacement, of distribution of weights, of the form of underbody and of topsides, are important and any design which violates this principle must fail of success. Every vessel is necessarily symmetrical transversely — that is, both sides are exactly alike — but longitudinal symmetry is not so necessary. Still, the forebody in its general bulk or mass should not be far different from the afterbody below and above the water. The freeboard forward will naturally be greater than at the stern, but enormously high bows and low sterns are distinctly bad for a variety of reasons. If the character of the design is such as to call for a wide, flat stern, the freeboard should not be as great as with a fine, sharp stern. It has the same bulk or volume on a smaller freeboard. Deckhouses and erections may be treated to carry out this idea of symmetry. An instance in point is the location of the funnel in steam yachts. This should be at, or close to mid-length to produce the best effect.

Still another principle the importance of which I wish to emphasize is *harmony*. A vessel is a complex creation having many features which, however perfect in themselves, must blend into a harmonious whole. Certain types of hull call for certain rigs or certain styles of deckhouses; certain types of bow belong with certain types of stern, for while each may be beautiful, considered separately, they may be utterly incongruous both for aesthetic and practical reasons when used together. The chief components of the visible entity of the boat are the bow, stern, contour of sheer, general freeboard, mass of deck erections, size and rake of spars, funnel, etc. It is impossible to lay down any test for harmony. Familiarity with this principle must come through practice and observation. A full appreciation of this subtle principle of harmony is evidenced only in the work of the most proficient designers. A man who has mastered it might be termed a naval artist.

PROGRESS IN SAILING YACHT DESIGN

Marked development in the form and rig of sailing yachts has taken place in the last thirty years. Overhanging bows, fin keels, longitudinal framing, hollow spars, crosscut sails and wire halliards are some of the great improvements and, remarkable to relate, practically all are the work of the great designer, Nathanael Greene Herreshoff.

The most marked development of form in sailing yachts which has occurred is in the shape of the bow. This is the natural outcome of racing rules which used water line length as a basis of comparison. It was found that sail-carrying power could be increased independently of water line length by increase of over all length, especially in the bow. This increasing of the stability is the principal function of overhangs, and results from an easing of sailing lines and a general shifting to leeward of the center of buoyancy when heeled. There is scarcely any limit to the increase in stability which may be secured by lengthening and flattening the overhangs. The tendency to overstep in this direction in racing yachts has, fortunately, been checked by rating rules, and in cruising yachts by the dictates of common sense.

The problem of designing a sailing yacht with speed as a foremost consideration is a most complex one. External conditions to which a yacht is subjected, such as force and direction of wind, condition of sea, etc., are constantly changing so that the attainment of a given speed may not be sought, but rather such a form as shall be easily driven at all speeds within appropriate limits. Nor is this the only consideration, for ease of form must be sacrificed to some extent for sail-carrying power. A harmonious adjustment between power and resistance should be sought. The problem of power yacht design is much more simple, as there the hull may be so designed as to be most easily driven at the desired speed, for, contrary to early theories, the hull most easily driven at one speed is not most easily driven at all speeds. Then, again, the power-driven yacht travels on an even keel, ordinarily, and the form of the yacht need be but little affected by considerations of stability.

STILL ROOM FOR IMPROVEMENT

In spite of the great advance made in yacht design in the past few decades, one need not fear that the possibilities are by any means exhausted. It should be the ambition of every designer to contribute his part to progress in the art. Eliminating from consideration the matter of improvement in marine motors, a field in which vast strides will probably be made within a few years, there are many directions in which advancement may be made — improved sails, possibly double surface sails with efficiently curved surfaces like aeroplane wings, improved spar stepping to eliminate depressing component of wind pressure, adjustable sparring for correction of balance, improved boat stowage, some simple anti-rolling device for power yachts, some method of temporarily increasing lifting surface area of hydroplanes when starting to plane, reduction of resistance of shaft supports, improved planing forms, etc.

The progressive designer will do a lot of experimental work on his own account for the accumulation of data along lines where it is scarce, such as for the determination of stresses in spars and rigging with various rigs, resistance of special forms, strengths of special light planking systems, wind pressures on sails, value of new rust-proofing processes and alloys, efficiency of bottom paints, holding power of various types of anchors, strength of turnbuckles, etc., etc.

Most of the investigations indicated may be made without expensive apparatus. Resistance, stresses, pressures and strength may be determined with sufficient accuracy to be instructive through the use of models in ways which I shall outline later. First hand information is the most valuable kind, and independent investigation in some of these fields will be of great value to the designer, directly and indirectly.

CHAPTER II

METHODS OF CALCULATION

A LARGE proportion of the calculations of the naval architect revert to the determination of the area and position of center of gravity of a figure bounded by straight lines and a curve. Two methods are in common use for these determinations, numerical and instrumental. The planimeter measures areas alone. It is an inexpensive instrument, and is practically indispensable to the yacht designer. The integrator is a more expensive instrument, and while it is not indispensable for small work it is a practical necessity for large work where the saving of labor resulting from its use is considerable. The integrator commonly measures areas and their moments about a given axis. Some

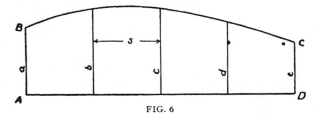

FIG. 6

integrators have, in addition, an attachment for measuring the moment of inertia of the area. Some of the uses of the planimeter and integrator will be explained later.

AREAS

The principal methods of obtaining areas numerically are by the use of Simpson's first or one-third rule, and by the trapezoidal rule. Referring to Fig. 6, let the area of the figure $ABCD$ be required. AD is a straight line, BC a fair curve, and a, b, c, d, e, are ordinates drawn perpendicular to AD at a distance, s, apart. The area of the figure is then, by Simpson's rule:

$$Area = 1/3s(a+4b+2c+4d+e)$$

This rule may be used with any odd number of ordinates. To apply the trapezoidal rule, simply add together all the ordinates, subtract one-half of the two end ordinates and multiply the sum by the common interval. Referring to Fig. 6 the area by the trapezoidal rule is:

$$Area = s(1/2a+b+c+d+1/2e)$$

If the bounding curve started at A and ended at D, making the end ordinates zero, the area would be:

$$Area = s(b+c+d)$$

This rule may be used with any number of ordinates.

This trapezoidal rule deals with the figure as though it were bounded

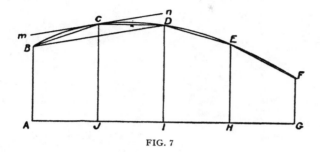

FIG. 7

by a broken line — that is, in Fig. 7, the area $ABFG$ by the trapezoidal rule is equal to the sum of the trapezoids $ABCJ$, $JCDI$, $IDEH$, and $HEFG$.

The area between BC, CD, DE, and EF and the curve is entirely neglected. This error is unimportant if the ordinates are sufficiently numerous, or if the curve is quite flat. Where there is a reverse in the curvature the loss on the convex portion of the curve is offset partially

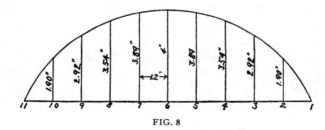

FIG. 8

or entirely by the gain on the concave portion. Simpson's rule assumes the portion of the curve BCD (Fig. 7) to be an arc of a parabola passing through C and tangent to mn (drawn parallel to BD), and the area BCD is added to, or subtracted from the area of the trapezoid $ABDI$, according as the curve is convex or concave.

For the purpose of comparing the two rules, let us apply them to the segment of the circle shown in Fig. 8. This curve, being entirely convex and steep at the ends, shows the trapezoidal rule at its worst. The height of the arc is four inches and the length of the chord twelve inches, making the interval 1.2 inches with eleven stations. The calculation is as follows:

STATION	ORDINATE	SIMPSON'S MULTIPLIERS	FUNCTIONS OF AREAS
1	0	1	0
2	1.90	4	7.60
3	2.92	2	5.84
4	3.54	4	14.16
5	3.89	2	7.78
6	4.00	4	16.00
7	3.89	2	7.78
8	3.54	4	14.16
9	2.92	2	5.84
10	1.90	4	7.60
11	0	1	0
Sum	28.50		86.76

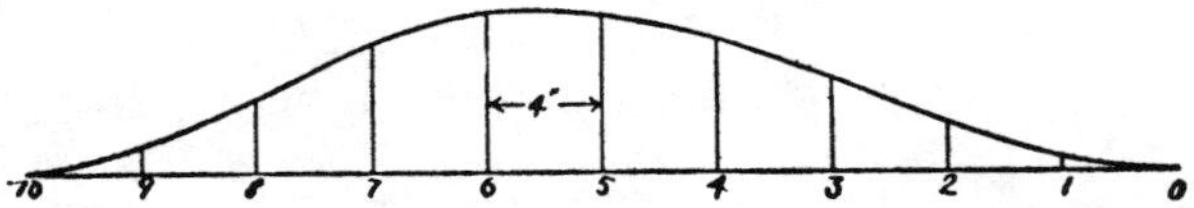

FIG. 9

Area by trapezoidal rule $= 28.50 \times 1.2 = 34.20$ square inches.

Area by Simpson's rule $= 86.76 \times \dfrac{1.2}{3} = 34.70$ square inches.

Knowing the radius of the arc, the correct area of the segment is readily found by direct computation to be 34.63 sq. in. The errors then amount to 1.2 per cent for the trapezoidal rule and 0.2 per cent for Simpson's. The latter is in excess of the correct amount while the former is below it.

AREA AND CENTER OF GRAVITY

To compare these rules still further, the area and center of gravity of the curve shown in Fig. 9 have been worked out by both rules.

This is the curve of areas of a yacht having a prismatic coefficient of .51. The common interval is 4 inches. To obtain the correct area, the trapezoidal rule, with a very large number of ordinates, was used, giving 121.82 square inches. This area is, of course, still approximate but the error is almost infinitesimal and may be disregarded. The calculation is as follows:

TRAPEZOIDAL   SIMPSON'S

STA.	ORD.	ARM.	MOM.	STA.	ORD.	SIMP'S. MULTS.	FUNCS. AREAS	ARM.	FUNCS. MOM.
0	0	5	0	0	0	1	0	5	0
1	.48	4	1.92	1	.48	4	1.92	4	7.68
2	1.71	3	5.13	2	1.71	2	3.42	3	10.26
3	3.36	2	6.72	3	3.36	4	13.44	2	26.88
4	4.85	1	4.85	4	4.85	2	9.70	1	9.70
			18.62						54.52
5	5.83	0	0	5	5.83	4	23.32	0	0
6	5.87	1	5.87	6	5.87	2	11.74	1	11.74
7	4.79	2	9.58	7	4.79	4	19.16	2	38.32
8	2.68	3	8.04	8	2.68	2	5.36	3	16.08
9	.83	4	3.32	9	.83	4	3.32	4	13.28
10	0	5	0	10	0	1	0	5	0
	30.40		26.81				91.38		79.42
			18.62						54.52
			8.19						24.90

Area by trapezoidal rule $= 30.40 \times 4 = 121.60$ square inches.

Area by Simpson's rule $= 91.38 \times 4/3 = 121.84$ square inches.

Center of gravity by trapezoidal rule $= \dfrac{8.19}{30.40} \times 4 = 1.08$ inches to the left of station 5.

Center of gravity by Simpson's rule $= \dfrac{24.90}{91.38} \times 4 = 1.09$ inches to the left of station 5.

The percentage errors in area are then, for the trapezoidal rule, .24 per cent; for Simpson's rule, .04 per cent. The difference in the positions of the center of gravity of the figure given by the two rules amounts to but .00025 of the base.

The process by which the center of gravity is found is briefly as follows: a station is selected near the middle of the length of base, in this case station 5, and each ordinate for the trapezoidal rule, or function of areas for Simpson's rule, is multiplied by the number of intervals between it and the station selected. These functions of moments are added separately on each side of the station about which moments are taken, the center of gravity being on the side of the larger sum. The difference of these sums divided by the sum of the ordinates for the trapezoidal rule, or by the sum of functions of areas for Simpson's rule, multiplied by the distance between stations, gives the distance from the center of gravity to the station about which moments were taken.

The comparisons just made of the working of the trapezoidal and Simpson's rule show the latter to be the more accurate. The trape-

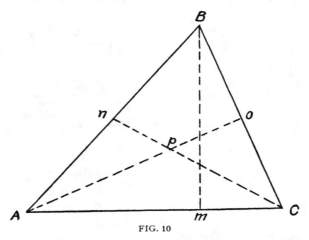

FIG. 10

zoidal rule, however, is quite accurate enough if a sufficient number of stations are taken (nine or more are recommended) and will be used for all calculations in the ensuing pages. It is superior to Simpson's rule

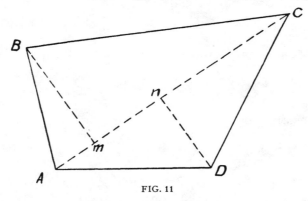

FIG. 11

in that it may be used with any number of stations and involves much less numerical work in its application. These considerations more than atone for its being slightly less accurate.

Some special calculations, which are frequently used in yacht designing, will now be presented.

The area of a triangle is equal to one-half the product of its base by its altitude; thus the area of the triangle ABC (Fig. 10) is equal to 1/2 $(AC \times Bm)$. The center of gravity of the triangle is at p, the point of intersection of Ao and Cn, n and o being the middle points of AB and BC respectively.

The area of a quadrilateral having no two sides parallel is found by

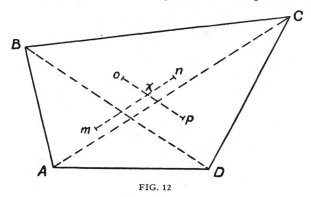

FIG. 12

dividing the figure into two triangles and finding the area of each separately. Thus the area of $ABCD$ (Fig. 11) is equal to 1/2 $(AC \times Bm) +$ 1/2 $(AC \times Dn)$. The center of gravity is on a line connecting the centers of the component triangles. In Fig. 12, m and n are the centers of the

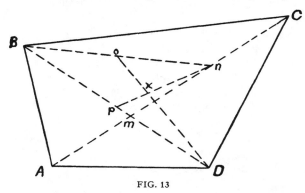

FIG. 13

triangles ABD and BCD, while o and p are the centers of the triangles ABC and ACD. Draw lines mn and op. Their point of intersection, x, is then the center of the figure $ABCD$. A somewhat quicker method of finding the center of such a figure is shown in Fig. 13. Draw the diagonals AC and BD. Lay off $Cn = Am$. Draw Bn. Bisect Bn and Bd, o and p being the middle points. Draw Do and np. Their point of intersection, x, is the center of gravity of the figure.

It is often necessary to find the common center of two areas or weights. In Fig. 14 let W and w represent two weights having their centers of gravity at a and b. Their common center lies at c on the line ab. Taking moments about a, the distance ac is equal to

$$\frac{w \times ab}{W + w}$$

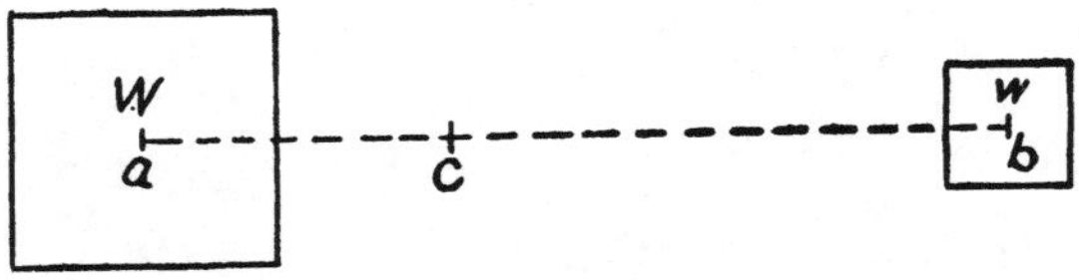

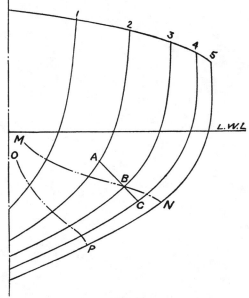

FIG. 14

WETTED SURFACE

There are various methods employed for finding the area of wetted surface of hull. These in general consist in applying Simpson's or the

FIG. 15

trapezoidal rule to the half girths below the water line, taken at each station. The difficulty lies in determining the correct interval, as the distance between stations on the surface of the skin is greater than at the center line, and is constantly varying. Taylor's mean secant method is probably the most accurate method for determining wetted surface.

In applying Taylor's method, each station is divided into a number of equal spaces below the water line, as Fig. 15 where the lines MN and OP divide the stations into thirds. At a point B on station 3, the line AC is drawn normal to the station and stopping at A on 2 and C on 4. Now let us pass a plane through AC perpendicular to the plane of the paper. Fig. 16 shows this plane revolved into the plane of the paper. $A'C$ is the projection of AC. If now we take a piece of paper and mark off on one edge of it the distance AA', which is twice the distance between stations, and then divide this distance decimally, we have a scale for measuring normals. If with this scale we measure AC, Fig. 15,

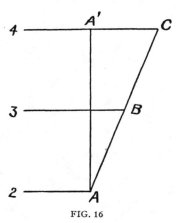

FIG. 16

TABLE I
NATURAL TANGENTS AND CORRESPONDING SECANTS

TANGENT	SECANT	TANGENT	SECANT	TANGENT	SECANT	TANGENT	SECANT
.010	1.000	.160	1.013	.310	1.047	.460	1.101
.020	1.000	.170	1.014	.320	1.050	.470	1.106
.030	1.000	.180	1.016	.330	1.053	.480	1.109
.040	1.001	.190	1.018	.340	1.056	.490	1.114
.050	1.001	.200	1.020	.350	1.060	500	1.118
.060	1.002	.210	1.022	.360	1.063	.510	1.123
.070	1.002	.220	1.024	.370	1.066	.520	1.127
.080	1.003	.230	1.026	.380	1.070	.530	1.132
.090	1.004	.240	1.028	.390	1.073	.540	1.137
.100	1.005	.250	1.031	.400	1.077	.550	1.141
.110	1.006	.260	1.033	.410	1.081	.560	1.146
.120	1.007	.270	1.036	.420	1.085	.570	1.151
.130	1.008	.280	1.038	.430	1.089	.580	1.156
.140	1.010	.290	1.041	.440	1.093	.590	1.161
.150	1.011	.300	1.044	.450	1.097	.600	1.166

($= A'C$, Fig. 16) we have at once the tangent of the angle $A'AC$. Now in a table of natural functions find the secant corresponding to this tangent. The secant gives the distance AC in terms of AA'. Next find the distance AC in the same manner at the other points on the station and find their mean.

Multiply the half girth by this mean secant. Treat the half girth at each station in the same manner and sum up by the trapezoidal rule, using for s the actual distance between stations. The result is a very close approximation to one-half the wetted surface. Table I gives tangents, advancing by hundredths, and the corresponding secants. The following calculation is for the wetted surface of the 30-footer (Fig. 57) by Taylor's method:

SECANTS

STATION	4	5	6	7	8	9	10	11	12
At L. W. L....	1.038	1.036	1.026	1.016	1.003	1.000	1.006	1.018	1.026
At M N......	1.033	1.028	1.016	1.007	1.001	1.000	1.007	1.024	1.031
At O P.......	1.031	1.025	1.016	1.008	1.001	1.001	1.011	1.006	1.028
At Keel.......	1.031	1.036	1.013	1.005	1.001	1.000	1.002	1.002	1.013
Sum.......	4.133	4.124	4.071	4.036	4.006	4.001	4.026	4.050	4.098
½ end ordinates	1.035	1.036	1.020	1.011	1.002	1.000	1.004	1.010	1.020
Difference.....	3.098	3.088	3.051	3.025	3.004	3.001	3.022	3.040	3.078
Divide by 3....	1.033	1.029	1.017	1.008	1.001	1.000	1.007	1.013	1.026
Half Girths....	1.54	2.78	4.23	5.63	6.05	6.26	6.42	6.26	4.35
Corrected Half Girths......	1.59	2.86	4.30	5.68	6.06	6.26	6.46	6.35	4.46

The corrected half girths are the products of the two preceding lines. The sum of the corrected half girths is 44.02. The wetted surface of the hull is then

$$44.02 \times 4/3 \times 2 \times 3 = 352.2 \text{ sq. ft.}$$

The factor 4/3 corrects for scale, the half girths having been measured in inches on the drawing. The factor 2 allows for both sides and 3 is the distance between stations. To the immersed area of hull we must add that of rudder and centerboard, both sides. These are, respectively, 24.8 and 25.6, making the total area of wetted surface 402.6 sq. ft.

The application of Taylor's method is somewhat laborious and for small work the bilge diagonal method will do well enough. This method consists in taking the half girths as before and applying the trapezoidal rule, using a corrected interval. This interval is obtained by multiplying the distance between stations by the length of a bilge diagonal between the stations at the forward and after points of immersion and then dividing by the length of the load water line. The corrected interval is generally about 3 to 4 per cent greater.

A fairly close preliminary approximation of the wetted surface may be made by use of the formula $S = C\sqrt{DL}$ where C is a coefficient varying according to type of boat. It is necessary to have worked out C for a number of designs similar to the one you are projecting. For motor yachts with faired after deadwood, C is about 16.

MOMENT OF INERTIA OF WATER LINE

To obtain the moment of inertia of a water line about its longitudinal axis, measure the half-breadths at each station, cube them, apply the trapezoidal rule to the cubes of ordinates and multiply the result by two-thirds. If the ordinates are measured in inches, as is generally most convenient, and the drawing is to other than the one-inch scale, the sum of the cubes must be multiplied by the cube of the inverted scale. The cubes of ordinates may be taken from a table of cubes of numbers. The calculation for the moment of inertia of the L. W. L., of the thirty-footer about a longitudinal axis is as follows:

STA.	ORD. INCHES	CUBES OF ORD.
4	1.43	2.91
5	2.52	16.00
6	3.31	36.26
7	3.91	59.78
8	4.29	78.95
9	4.37	83.45
10	4.19	73.56
11	3.74	52.31
12	2.88	23.89
		427.12

$$I = 427.12 \times 2/3 \times 3 \times 4/3^3 = 2025.$$

The factor 3 is the distance between stations and $(4/3)^3$ corrects for scale, the ordinates having been measured in inches.

To find the moment of inertia of a water line about a transverse axis through its center of gravity, we must first find the area of the water line and the position of its center of gravity. We may then find the moment of inertia of the water line about the midship section and make a correction to obtain the same about the center of gravity of the water line. Referring to the following computation on the 30-footer we have, in the second column, the half-breadths of the load water line plane measured in actual inches on the drawing.

STA.	ORD.	ARM.	FUNC. FOR C. G. OF L. W. L.	ARM.	FUNC. FOR I
4	1.43	5	7.15	5	35.75
5	2.52	4	10.08	4	40.32
6	3.31	3	9.93	3	29.79
7	3.91	2	7.82	2	15.64
8	4.29	1	4.29	1	4.29
			39.27		
9	4.37	0	0	0	0
10	4.19	1	4.19	1	4.19
11	3.74	2	7.48	2	14.96
12	2.88	3	8.64	3	25.92
	30.64		20.31		170.86

The third column gives the number of spaces each ordinate is from the midship section. The product of the second and third columns gives the functions for center of gravity of water line. The area of the water line is $30.64 \times 2 \times 4/3 \times 3 = 245.1$ sq. ft. The factor 2 allows for both sides, 4/3 corrects for scale and 3 is the distance between stations. The center of gravity is

$$\frac{39.27 - 20.31}{30.64} \times 3 = 1.86 \text{ ft. forward of sta. 9.}$$

Multiplying the functions for center of gravity again by the lever arms, we get the functions for moment of inertia. These are all added together and for the moment of inertia about station 9 we get $170.86 \times 2 \times 4/3 \times 3 \times 3^2 = 12,302$.

The factor 2 allows for both sides, 4/3 corrects for scale, 3 is the interval and 3^2 corrects for the lever arms, being in each case the number of intervals instead of the actual distance in feet between each ordinate and the midship section. Now, to find the moment of inertia about the center of gravity of the water line we must subtract the area of the water line times the square of the distance between its center of gravity and the midship section. This correction is $245.1 \times 1.86^2 = 848$. The moment of inertia about the center of gravity is then $12,302 - 848 = 11,454$.

WEIGHT CALCULATION

The calculation of the weight of the yacht is of great importance where it is necessary to produce exact flotation on a given *L.W.L.* and the designer is not in possession of accurate weight data on a very similar boat. The weight of each member should be calculated or estimated separately, the process for structural members being to find the volume and multiply by the density.

Table II gives average values for the densities of various materials entering into the structure and equipment of the yacht. The volume of

TABLE II

WEIGHT OF A CUBIC FOOT OF SUBSTANCES

METALS	CAST	ROLLED
Aluminum and Duralumin..........................	165	165
Brass...	534	534
Bronze...	509	
Bronze, Manganese...............................	530	
Bronze, Tobin....................................		525
Copper...	556	556
Iron...	442	485
Lead...	710	
Monel Metal......................................	552	558
Steel..		487

LIQUIDS	LBS. PER CU. FT.	LBS. PER GAL
Alcohol..	50.0	6.7
Gasoline...	45.7	6.1
Kerosene...	50.9	6.8
Oil, Boiler......................................	56.0	7.5
Oil, Diesel......................................	53.0	7.1
Oil, Linseed.....................................	58.0	7.8
Oil, Lubricating.................................	58.2	7.8
Red lead paint...................................	210.0	28.0
Turpentine.......................................	54.0	7.2
Water, Fresh.....................................	62.4	8.3
Water, Salt......................................	64.0	8.6

7.48 gals. = 1 cu. ft.

WOODS	WEIGHT PER CU. FT. SEASONED	GREEN
Ash, white.......................................	40	49
Balsa..	9	
Birch, yellow....................................	32	57
Butternut..	35	45
Cedar, Oregon....................................	31	39
Cedar, Red or White..............................	22	41
Cedar, Spanish...................................	35	42
Elm, Rock..	45	48
Fir, Douglas.....................................	32	34
Hackmatack.......................................	35	
Hickory..	48	
Locust...	46	66
Mahogany, Honduras...............................	33	39
Mahogany, Philippine.............................	36	39

TABLE II — *Continued*

Maple, Sugar	43	46
Oak, White	46	62
Pine, Oregon	32	40
Pine, White	26	39
Pine, Yellow	44	47
Spruce	27	28
Teak	48	58
Walnut	42	

MISCELLANEOUS

Boiler Punchings	350
Cement	183
Cement, Stone and Sand	144
Cement and Cinder	100
Coal, Hard	47
Coal, Soft	40
Glass, plate	161
Rubber, dense	98
Rubber, sponge	36

the planking is found by multiplying its area in square feet by its thickness in feet. The area is found in the same manner as the wetted surface, the half girths being taken from planksheer to rabbet line. The volume of the frames is found by multiplying the area of surface next the planking by the mean moulded depth. The area of outer surface depends on the siding and spacing; thus if the frames were sided 2 inches and spaced 14 inches the area would be $2/12 \times 12/14 = 1/7$ that of the planking. The volume of the deck is found by multiplying its area, which is readily found, by its thickness. The volume of the beams is found in the same manner as that of the frames. The volume of the deadwood is found by applying the trapezoidal rule to the area of transverse sections taken at regular intervals. The volumes of long members, like stringers, clamps, spars, etc., are equal to the length multiplied by the mean sectional area.

Table III gives the weights of spruce spars of various diameters. The most convenient methods of obtaining the weights of other portions of hull, rig and equipment will be apparent. Tables IV and V give weights of sails and fastenings.

Care must be taken that nothing is omitted from the schedule of weights. There are some things such as paint, cabin fittings, personal effects, etc., which elude calculation and whose weight must be estimated. A fair allowance for outside paint is nine pounds per hundred square feet of surface over plain wood, nineteen pounds for canvas. The calculation of weights for the thirty-footer is given on page 68.

TABLE III

WEIGHTS OF SOLID SPRUCE SPARS PER FOOT OF LENGTH

(For Oregon Pine Multiply by 1.13)

DIAM.	WT.	DIAM.	WT.	DIAM.	WT.	DIAM.	WT.	DIAM.	WT.
3″	1.62	5¼	4.96	7½	10.10	9¾	17.1	14	35.3
3¼	1.90	5½	5.43	7¾	10.8	10	18.0	14½	37.8
3½	2.20	5¾	5.95	8	11.5	10½	19.8	15	40.5
3¾	2.53	6	6.54	8¼	12.2	11	21.8	15½	43.2
4	2.88	6¼	7.02	8½	13.0	11½	23.8	16	46.0
4¼	3.25	6½	7.60	8¾	13.8	12	25.9	16½	48.8
4½	3.65	6¾	8.20	9	14.6	12½	28.1	17	52.0
4¾	4.07	7	8.82	9¼	15.4	13	30.2	17½	55.0
5	4.50	7¼	9.46	9½	16.2	13½	32.7	18	58.3

TABLE IV

WEIGHTS OF SAILS IN POUNDS PER 100 SQ. FT. OF AREA

Cloths 16″ Wide

NO.	WT.
000	26.2
00	24.9
0	23.7
1	22.5
2	21.2
3	20.0
4	18.8
5	17.5
6	16.2
7	15.0
8	13.8
9	12.6
10	11.4

Cloths 20″ Wide

OZ.	WT.
12	14.1
10	12.6
8	11.0

Cloths 28½″ Wide

7	6.9
6	6.0
5	5.0
4	4.0

TABLE V

FASTENINGS

NUMBER OF NAILS IN ONE POUND

LENGTH	COPPER NAILS	GALV. BOAT NAILS	GALV. WIRE NAILS
¾″	710	500	800
⅞″	600	450	700
1″	495	400	550
1¼″	320	300	400
1½″	215	200	260
1¾″	155	135	170
2″	115	95	110
2¼″	85	68	80
2½″	65	50	70
3″	44	33	57
3½″	30	24	43
4″	20	20	30

The accuracy of weight calculations on wooden boats is vitiated by the variation in densities of woods but, with care, fairly close results may be obtained. The weight of green timber should be used for parts liable to soakage. This difficulty does not exist with metal construction as the densities of plates and shapes are accurately known and, of course, do not vary appreciably.

The plating of metal yachts is not of uniform thickness but is made heavier in localities which are especially stressed. For this reason the skin cannot be dealt with as a whole but the weight of each plate or group of plates of the same weight must be figured separately. The dimensions of the plates should be taken from the shell expansion. The length, however, is not shown correctly in the expansion and a correction should be made by multiplying by the secant of the angle which a diagonal through the given plates makes with the center line. The weight of plates is expressed in pounds per square foot, and a 40-pound steel plate is considered to be an inch thick, making five pounds to the eighth of an inch. A bronze plate an inch thick weighs 43 pounds per square foot.

The length of frames and reverse frames should be measured on the body plan, making allowance for clips and doublings. The sum of the lengths of frames of the same size is to be multiplied by the weight per foot to get their weight. Weights of keels, keelsons, stringers and other longitudinal members may be found from their length, allowing for curvature, multiplied by their weight per foot. The weight of floors, deck stringer plates, bracket and tie plates may be calculated from their area. In finding the weight of deck beams, their length should be measured along the upper crowned side and allowance should be made for the part turned down to form the bracket, if this construction is used. The weights of vertical laps, butt straps, liners under outer strakes, and rivet heads may be computed, but the calculation would be extremely laborious, and it is customary to allow for these items by arbitrary percentages taken from practice. A fair allowance for these items is — butts, liners and rivets in plating, 10 per cent; rivets in frames and reverse frames, 5 per cent; rivets in floors and brackets, 3 per cent.

Small yachts should be weighed after completion whenever possible as a check on the calculations. Naval vessels are often weighed during construction, that is, the weight of all the material worked into the ship is carefully recorded as well as the weight of all refuse taken out. This procedure might be employed to advantage in important yacht work.

Calculations for longitudinal strength and of stresses set up when among waves are sometimes made. As these calculations are most complex and are made only on large or lightly built steamers, it is thought unnecessary to give space to them here and the reader is referred to works on general naval architecture for these calculations.

Figure 17 gives formulæ for computing suitable size of free-end

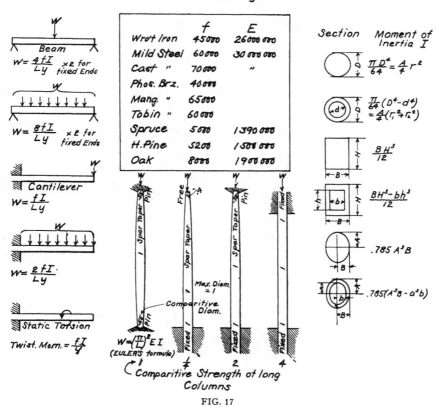

Strength of Beams, Shafts and long Columns

W= Breaking Load, lbs.   f = max. fibre Stress, lbs.   L = Length, Inches
I = Mom. of Inertia of Section   y = distance extreme Fibre from neutral Axis
E = Modulus of Elasticity

Section	f	E
Wro't Iron	45000	26000000
Mild Steel	60000	30000000
Cast "	70000	"
Phos. Brz.	40000	
Mang. "	65000	
Tobin "	60000	
Spruce	5000	1390000
H. Pine	5200	1500000
Oak	8000	1900000

Beam
$$W = \frac{4fI}{Ly} \quad \times 2 \text{ for fixed Ends}$$

$$W = \frac{8fI}{Ly} \quad \times 2 \text{ for fixed Ends}$$

Cantilever
$$W = \frac{fI}{Ly}$$

$$W = \frac{2fI}{Ly}$$

Static Torsion
$$\text{Twist. Mom.} = \frac{fI}{y}$$

$$W = \left(\frac{\pi}{L}\right)^2 EI \quad \text{(EULER'S formula)}$$

Section   Moment of Inertia I

$$\frac{\pi D^4}{64} = \frac{A}{4} r^2$$

$$\frac{\pi}{64}(D^4 - d^4) = \frac{A}{4}(r_1^2 + r_2^2)$$

$$\frac{BH^3}{12}$$

$$\frac{BH^3 - bh^3}{12}$$

$$.785 A^3 B$$

$$.785(A^3B - a^2b)$$

Max. Diam. = 1
Comparitive Diam.

Comparitive Strength of long Columns

FIG. 17

beams, cantilevers, pillars and shafts. f in the formulæ is the maximum fiber stress and the member should be of such a size that this stress is not greater than the load the member is to bear, multiplied by a factor of safety; 4 is a good average value for the safety factor. The beam formulæ are used for computing the size of specially loaded deck beams, booms, engine room crane beams, boat transfer beams, etc, The cantilever formulæ are useful for figuring strength on unsupported masts, bowsprits, davits, etc. Fig. 18 illustrates this application to ordinary davits. The long column formula is useful for figuring masts, and special application of it to spruce masts is explained in Chapter

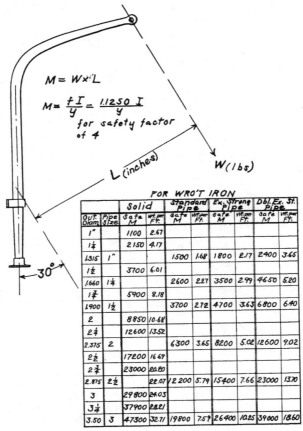

$$M = W x \cdot L$$

$$M = \frac{f I}{y} = \frac{1.1250\, I}{y}$$
for safety factor of 4

L (inches) W (lbs)

FOR WRO'T IRON

Solid				Standard Pipe		Ex. Strong Pipe		Dbl. Ex. St. Pipe	
Out. Diam.	Pipe Size	Safe M	Wt. per Ft.	Safe M	Wt. per Ft.	Safe M	Wt. per Ft.	Safe M	Wt. per Ft.
1"		1100	2.67						
1¼		2150	4.17						
1.315	1"			1500	1.68	1800	2.17	2400	3.65
1½		3700	6.01						
1.660	1¼			2600	2.27	3500	2.99	4650	5.20
1¾		5900	8.18						
1.900	1½			3700	2.72	4700	3.63	6800	6.40
2		8850	10.68						
2¼		12600	13.52						
2.375	2			6300	3.65	8200	5.02	12600	9.02
2½		17200	16.69						
2¾		23000	20.80						
2.875	2½		22.07	12200	5.79	15400	7.66	23000	13.70
3		29800	24.03						
3¼		37900	28.21						
3.50	3	47300	32.71	19800	7.57	26400	10.25	39000	18.60

FIG. 18

XI. It is also useful for figuring stanchions, spreaders, booms, struts and other members in direct compression. The static torsion formula is useful for rudder posts, winch shafts and other members of very low speed of rotation. I is the polar moment of inertia.

DAVITS

In calculating size of davits, it is well to assume the yacht heeled as shown to 30° to allow for the worst conditions, also to figure in water or crew in boat if these extra weights are likely ever to be hoisted with boat. Suppose boat and contents weigh 500 lbs. and heeled arm L is 50 inches. Then, bending moment $M = 250 \times 50 = 12,500$ inch lbs. Turning to the table, Fig. 18, we find that this bending moment is safe for $2\frac{1}{4}''$ solid wrought iron or $2\frac{1}{2}''$ standard pipe or $2''$ double extra strong pipe. The maximum diameter is required at the upper bearing.

TO CONSTRUCT AN ELLIPSE

The designer has frequent occasion to construct an ellipse, as, for instance, in drawing port lights, manholes, developed propeller blades, etc. Fig. 19 shows a simple method. Knowing the semi major axis $O\,A$

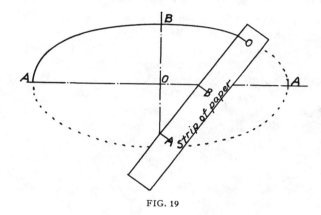

FIG. 19

and semi minor axis $O\,B$, lay these distances off on a strip of paper. Then, keeping point B on the strip on the major axis and point A on the minor axis, dot a series of points opposite O on the strip.

In conclusion it may be said that the designer should avoid long hand figuring for computation, using logarithms or calculating machine exclusively. Otherwise, any extended calculations is so laborious as to render the work unprofitable. The ordinary slide rule, especially the 20 inch length, is quite good enough for a large portion of the calculations on small yachts. Logarithms leave the work in a much more convenient form for reference than ordinary figuring and are especially valuable for finding powers and roots of numbers. In addition, they are more expeditious and more accurate than ordinary figuring. The use of logarithms is not difficult and should be mastered, as they afford the only means of raising numbers to fractional powers — as is necessary in powering and propeller calculations.

CHAPTER III

RELATIVITY AND SIMILITUDE

R ELATIVITY is a word which has been used a great deal of late by
scientists in connection with abstruse theories of physical and
astronomical phenomena. This use of the word is little understood by
anyone but scientists. It is a good word to use, however, in connection
with acquiring a conception of the characteristics of boats and their re-
lationship in the matters of dimensions, speed, stability, etc.

NAUTICAL RELATIVITY

In judging as to whether a vessel is deep or shallow, heavy or light,
fast or slow, some standard of comparison is necessary as all these
qualities are relative. The characteristic most easily visualized and best
adapted to use as a basis of comparison is the length on the load water
line, and to this most of the other features are compared by use of
various ratios. Many of these are set forth in the table on page 31. The
various abbreviations or symbols noted are commonly used and will be
employed elsewhere throughout the book.

The use of these ratios establishes the normality, or otherwise, of any
dimension. Thus, if the beam of a boat is given as 10 feet one does not
know whether this wide or narrow until it is compared with the
length, and then the ratio of length to beam tells the story at once, taking into consideration the type.

In the same way, speed is relative and figures mean nothing until
length and type are taken into consideration. In general, for heavy
displacements, speeds are considered low when speed-length ratio is
less than .7 and high when it is over 1.3. *Leviathan* is thought of as a
fast ship, but when one considers that her speed-length ratio is only
about .9 he realizes that she is slow — considering her size. It would take
over a million horse power to bring her speed-length ratio up to 1.5.

Fig. 20 gives an interesting comparison of the speeds of various
types of vessels based on their speed-length ratios. Also a comparison

of their powering by the scale showing ratios $\dfrac{P}{D^{7/6}}$, from which com-

parison planing types are excluded. Average displacement-length
ratios for corresponding speeds are also shown.

There is a wide variation in displacement-length ratios with size and
type. In the case of power yachts, extremely high speeds (still excluding

TABLE VI

FEATURE	SYMBOL	RATIO FOR COMPARISON	EXPRESSION
Length on L.W.L.	L	Base	
Beam, extreme	B	Ratio length to Beam	$\dfrac{L}{B}$
Draft, extreme	H	Ratio Beam to Draft	$\dfrac{B}{H}$
Area Midship Section	MS		
Area Wetted Surface	S	S to MS or $D^{2/3}$	$\dfrac{SA}{WS}$ or $\dfrac{S}{D^{2/3}}$
Area Sails	SA	SA to L^2 or MS or S	
Fineness		Prismatic Coefficient	$\dfrac{35D}{L \times MS}$
Fineness		Block Coefficient	$\dfrac{35D}{L \times B \times H}$
Displacement in tons of 2240 lbs.	D	Displacement-Length Ratio	$\dfrac{D}{\left(\dfrac{L}{100}\right)^3}$
Speed in Knots	V	Speed-Length Ratio	$\dfrac{V}{\sqrt{L}}$
Resistance, frictional, lbs.	R_f	Lbs. per ton	$\dfrac{R_f}{D}$
Resistance, residual	R_w	Lbs. per ton	$\dfrac{R_w}{D}$
Effective Horse Power	$E_t = E_f + E_w$	hp. per ton to 7/6 power	$\dfrac{E_t}{D^{7/6}}$
Brake Horse Power	$P = P_f + P_w$	hp. per ton to 7/6 power	$\dfrac{P}{D^{7/6}}$
Propulsive Efficiency		Ratio effective to brake hp.	$\dfrac{E_t}{P_t}$

planing types) are associated with displacement-length ratios down to 40, and in the other direction up to 150 for slow speed boats of houseboat type. A conception of a boat's displacement will always be best gained through the use of this ratio.

A rough and ready method of approximating the power necessary to drive a boat at a given speed is to pick out the ratio $\dfrac{P}{D^{7/6}}$ for the given speed-length from Fig. 20. Then this ratio times $D^{7/6} = P$. Values of

$D^{7/6}$ may be taken quickly with sufficient exactness for the purpose from Figs. 21 and 22.

This and many other curves throughout the book have ordinates plotted logarithmically for greater accuracy in the lower values.

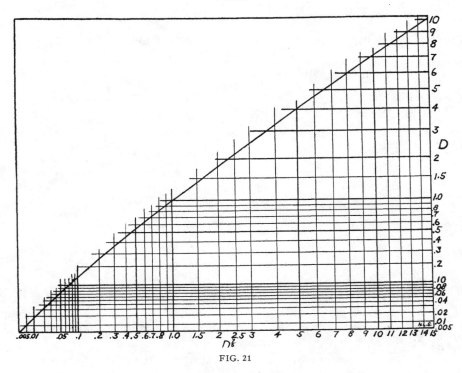

NAVAL and COMMERCIAL	AUX. SHIPS	TRAMPS TANKERS	FAST FREIGHTERS	MAIL SHIPS	BATTLESHIPS	BATTLE CRUISERS	CHANNEL STRS.	SCOUT CRUISERS	DESTROYERS				TURBINIA AND ARROW			
SPEED	LOW		MODERATE	MEDIUM			HIGH SPEED				VERY HIGH SPEED						
$D \div (\frac{L}{100})^3$				90 80	70	60	50			40							
$V \div \sqrt{L}$	.4	.6	.8	1.0	1.2	1.4	1.6	1.8	2.0	2.2	2.4	2.6	2.8	3.0	3.2	3.4	3.6	3.8
$P \div D^{\frac{2}{3}}$			.1	.2	.4 .6 .8	1	2	3	4 5		10	15	20	25	30	35	40	45
YACHTS	AUXILIARIES & HOUSE BOATS		FULL POWERED CRUISERS		FAST CRUISERS & RUNABOUTS Round Bottom						VERY FAST CRUISERS AND RUNABOUTS V Bottom						

FIG. 20

FIG. 21

SIMILITUDE

When vessels are of exactly similar form but of different size their various characteristics vary in accordance with the *law of mechanical similitude*, also known as Froude's law of Comparison. Some of the most important expressions of this Law are as follows:

Beam, draft or any other lineal dimension varies as L and $D^{1/3}$.

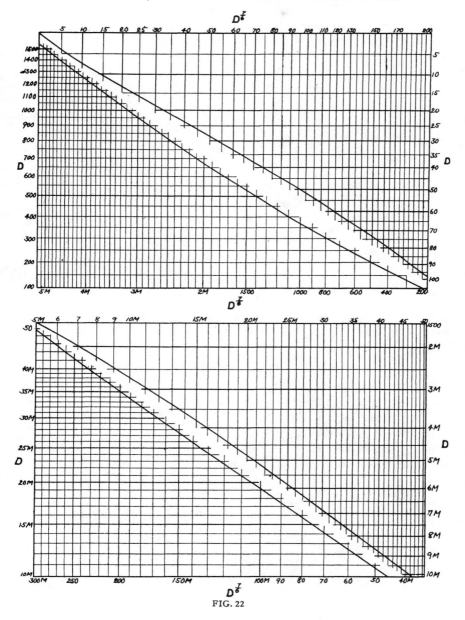

FIG. 22

Sail area, wetted surface or any other area varies as L^2 and $D^{2/3}$.

Displacement, ballast or any other weight varies as L^3 or $S^{3/2}$.

Stability varies as L^4.

Heeling moment of wind pressure on sails varies as L^3.

Moment of inertia varies as L^5.

Resistance varies as L^3 and D.

Speed varies as $L^{1/2}$ and $D^{1/6}$.

Horse power varies as $L^{7/2}$ and $D^{7/6}$.

The latter ratio does not hold exactly, for reasons which will be explained later.

In accordance with the above, if one vessel has twice the length of another similar vessel, she will have:

> twice the beam
> four times the area of sail and wetted surface
> eight times the displacement
> eight times the resistance
> 1.41 times the speed $(1.41 = \sqrt{2} = 2^{1/2})$
> 11.3 " " horse power $(11.3 = 2^{7/2} = 8^{7/6})$
> 16 " " stability
> 8 " " heeling moment.

From the last two lines we realize that stability, or power to carry sail, increases much faster than the heeling moment due to wind pressure, which is the reason in a nutshell why large sailing yachts are so much stiffer than small ones with relatively much less draft and displacement.

PARTIAL SIMILITUDE

It is frequently desirable to reproduce a design with some variation, such as with all breadths increased in a certain ratio, or the breadth and depth or length only. Any one or two of the three principal dimensions may be varied and the resulting form will be similar in a way so that the displacement and surface area may be ascertained in advance by the law of *partial similitude*.

With B or H changed	D varies as B or H
" L changed	D " " L
" " "	S " " L (nearly)
" B and H changed	D " " B^2
" " " " "	S " " B
" " " " "	S " " $\sqrt{D}$

Variations in horse power and speed under above conditions will be treated in the chapter on resistance.

CHAPTER IV

DISPLACEMENT

A NY object floating in water displaces its own weight, that is, if the object were placed in a vessel brim full of water, an amount would run over equal to the weight of the object. Displacement of yachts up to

FISHERMAN TYPE SCHOONER YACHTS

35 or 40 feet water line is commonly expressed in pounds and above that size in tons of 2,240 pounds. The displacement is determined from the lines of the yacht, as will be explained later. There are approximately 35 cubic feet of salt water and 36 cubic feet of fresh water in one ton. In one cubic foot there are 64 pounds of salt water and 62.4 pounds of fresh water.

CENTER OF BUOYANCY

The center of buoyancy is the center of volume of the submerged portion of the vessel. The upward force of the buoyancy of the water may be considered to act at this point, and the weight of the vessel to act at the center of gravity. The centers of buoyancy and gravity must be in the same vertical line or the vessel will alter her trim so as to bring them so.

In a yacht it is important to know the longitudinal position of the center of buoyancy when the yacht is erect, to be able to distribute the weights so that she will trim properly. The vertical position of the center of buoyancy is not so important, although a knowledge of its position is useful, as will be shown later. Its transverse position is, of course, in the central vertical plane.

CURVE OF AREAS

If we plot the areas of the various transverse sections on a base line, as in Fig. 8, we have what is known as the curve of areas. The area under this curve gives us the displacement of the yacht, and the center of gravity of this area, projected on the base line, gives us the longitudinal position of the center of buoyancy. The process of determining displacement and longitudinal position of the center of buoyancy is best explained by a concrete example, the calculation for the 30-footer shown in Fig. 57 being as follows. The quantities opposite the station numbers are the planimeter readings of each half section in square inches. The distance between stations is 3 feet and the scale of the drawing three quarters of an inch to the foot.

DISPLACEMENT

STA.	½ A. SQ. IN.	ARM. .	MOM.
3	0		
4	.45	5	2.25
5	1.65	4	6.60
6	3.48	3	10.44
7	5.41	2	10.82
8	6.62	1	6.62
			36.73
9	6.71	0	
10	5.80	1	5.80
11	3.79	2	7.58
12	1.55	3	4.65
13	0		
	35.46		18.03

Displacement $= 35.46 \times 2 \times 16/9 \times 3 \times 64 = 24{,}207$ lbs.

$CB = \dfrac{36.73 - 18.03}{35.46} \times 3 = 1.58'$ for'd station $9 = 16.42'$ abaft sta. 3.

The factor 2 is for both sides of the yacht, 16/9 corrects for scale, being the square of the inverted scale, $(4/3)^2$. The lines are drawn on the scale $3/4'' = 1'$ and the planimeter reads in square inches, so it is necessary to correct its readings to give square feet on the boat. 3 is the distance between stations required in the application of the trapezoidal rule, and 64 is weight of one cubic foot of salt water.

The vertical position of the center of buoyancy may be found by taking the areas of water lines spaced at equal intervals from the load line down and summing up their moments in the same manner as for the longitudinal position. The vertical position is more readily found by the integrator if one be available, using the load line as an axis. The use of the integrator is described in Chapter V.

DETERMINATION OF DISPLACEMENT

The determination of the proper displacement of a yacht is the most important step in laying out a design. The displacement must equal the sum of the weights of the various members of the yacht and its contents or the yacht will not float as designed. The logical procedure, then, is to ascertain the component weights and make the displacement equal to their sum. For a design where the limitations as to the water line length and amount of ballast are stringent, the best procedure is to lay out a preliminary design which shall have the proper displacement and general dimensions as nearly as it is possible to estimate them, and to then make calculations for weight on this design as outlined in Chapter II. The displacement for the final design will then be made equal to the sum of these weights. For cruising boats it does well enough to estimate the proper displacement from data on successful boats of similar type and, after calculating the weight, to assign an amount of ballast to supply the deficiency of weight. If the proper displacement is known fairly close and a portion, or all, of the ballast is to be carried inside, the calculation for weight may be dispensed with as the amount of weight is easily adjusted to secure the desired flotation. The curves of displacement in Figs. 23, 24 and 25 will serve as guides in estimating displacement for sailing yachts.

Fig. 23 is for displacements 10–30 ft. *L.W.L.*
"   24 " "   "   30–60
"   25 " "   "   60–120

DISPLACEMENT FROM CURVE

The upper curves in Figs. 23, 24 and 25 give suitable displacements for normal sailing cruisers; that is, yachts having beam, draft, sail area

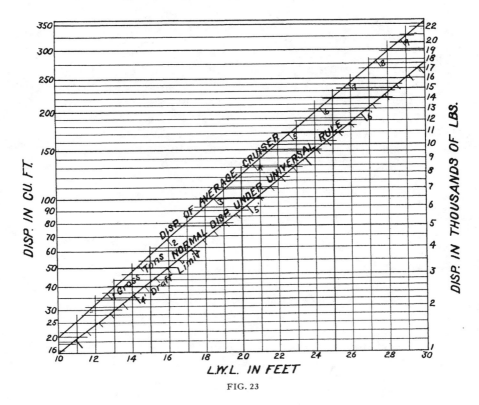

FIG. 23

and freeboard about as indicated on the other curves showing those dimensions. It may be said that many American cruisers, to their detriment, are designed with less displacement than these curves indicate. Ample displacement insures internal room, sail carrying power and seagoing ability — all prime requisites of the cruiser — and with proper designing the speed under most conditions is as good as with skimped displacement. Many yachts have more displacement than they were designed to have by going below their designed L.W.L., which is bad designing. Cruisers with extra heavy hulls or with a lot of top weight in equipment may well have even more displacement than is indicated on these curves.

The lower curves in Figs. 23, 24 and 25 indicate the normal displace-

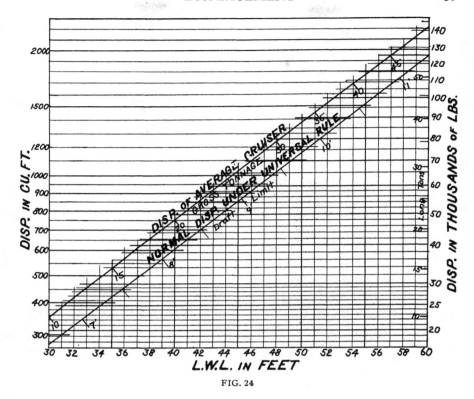

FIG. 24

ments called for by the Universal Rule which is explained in Chapter XIII.

The best position for the center of buoyancy is something of which little is known. An examination of the data for a large number of representative existing sailing yachts shows its position to lie generally between 50 to 56 per cent of the load water line length from its forward end. From 52 to 54 per cent is a suitable position for cruising boats.

PRISMATIC COEFFICIENT

The fineness of a design is commonly represented by three coefficients. These are the midship section coefficient or ratio of midship section to the circumscribed rectangle; the block coefficient or ratio of the volume of the displacement to the volume of the circumscribed parallelepipedon; and the prismatic coefficient or ratio of the volume of the displacement to the volume of a solid whose length is equal to the length of the water line and having a constant sectional area equal to that of the midship section. It is also the ratio of the area of the curve

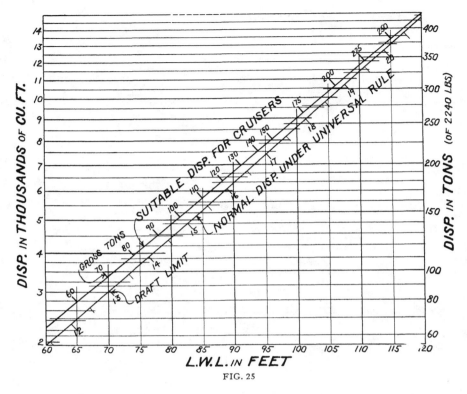

FIG. 25

of the transverse areas to the area of the circumscribed rectangle. The last coefficient is the most important for our purposes. The area of midship section of the thirty-footer is

$$6.80 \times 2 \times 16/9 = 24.2 \text{ sq. ft.,}$$

and the prismatic coefficient is equal to

$$\frac{24207}{30 \times 24.2 \times 64} = .522.$$

For power yachts, the prismatic coefficient generally lies between .54 and .60. For centerboard boats having little external keel, it varies from .53 to .61. For boats of the semi-keel type, from .50 to .54. For the modern keel boat, it varies from .46 to .54. Having decided on a suitable prismatic coefficient for a design, the area of the midship section is readily calculated for a given displacement as follows:

$$A = \frac{D \text{ cu. ft.}}{LWL \times PC}$$

The form of the curve of areas is a matter of considerable signifi-
cance, as it shows the manner in which the displacement is distributed
longitudinally. A design may be perfectly fair and sweet yet have an
undesirable distribution of the displacement as revealed by the area
curve. The form of the curve has undoubtedly great influence on the
wave-making resistance and should, of course, be that of least resistance for a given displacement and speed. According to the wave-form
theory as proposed by Colin Archer in 1877, the curve should take the
form of a wave line. A complete discussion of the wave-form theory
would not accord with the purpose of this book, and the application of
the theory to yacht design will simply be given. The wave form theory
requires the curve of areas to be a curve of versed sines for the forebody,
or .6 of the *L.W.L.*, and a trochoid for the afterbody. The length of
forebody in well-formed yachts is generally between 54 and 58 per cent
of the water line length, that is, the midship section or point of greatest
sectional area is in the neighborhood of 56 per cent of the water line
length from the forward end.

Before proceeding further it will be well to give the construction for
the curve of versed sines and the trochoid. Suppose, in Fig. 26, it is desired to construct a curve of versed sines of length *AC*. Let *AB* be the
diameter of generating circle. Divide the semi-circumference into a

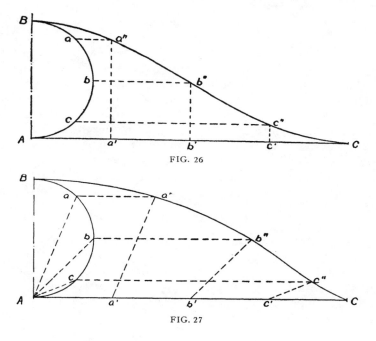

FIG. 26

FIG. 27

number of equal divisions. Divide the base line, *AC*, into the same number of equal divisions. Draw lines parallel to *AC* through *a*, *b*, *c*, and perpendiculars through *a'*, *b'*, *c'*. The points of intersection *a''*, *b''*, *c''* lie on the curve. Fig. 27 shows the construction for the trochoid. The semicircumference and base line are spaced off as before. Draw the chords *Aa*, *Ab*, and *Ac*. Draw *a' a''*, *b' b''*, *c' c''*, parallel and equal to the chords *Aa*, *Ab*, and *Ac*. The points *a''*, *b''*, and *c''* lie on the curve. The area of the versed sine curve is always equal to one-half the product of base line by diameter of generating circle, that is, its coefficient is .5. The trochoid has a varying coefficient dependent on the ratio of diameter of generating circle to base line, the coefficient increasing with the ratio. The height of the trochoid may be increased to pass through any desired point, having first drawn it with the proper diameter of generating circle to give the desired coefficient, by multiplying each ordinate by the height at the point through which the curve is to pass, divided by the ordinate of the original curve at the same point.

The curve of versed sines, the construction of which is explained above, is much used for area curves generally with the fine termination trimmed or "snubbed" as it is called. In this way the prismatic coefficient is increased more or less according to the amount of snubbing and the curve rendered much more suitable for a curve of areas.

There is undoubtedly an advantage in having the area curve conform to the wave-form theory. Its importance, however, should not be overestimated, as this in itself will not produce a speedy design. On the contrary, many successful yachts have a curve differing widely from that prescribed by the wave-form theory. The area curves of centerboard boats having little external keel may be made to take an exact wave form, decreasing, however, the length of forebody from .60 of the water line length to .56 or .57 to accord with modern practice. In boats having a pronounced external keel, it is desirable to treat boat and keel separately, making the areas of the boat proper take the wave form. Table VII gives a series of factors for constructive wave-form curves having various prismatic coefficients.

AREA CURVE FOR SAILING YACHTS

To construct an area curve using these factors, divide the base line between the forward and after points of immersion into ten equal parts. Erect ordinates at these points and, having selected a suitable prismatic coefficient, multiply the area of midship section by the factors of the curve having the coefficient and plot these quantities on the ordinates. A curve through these points will have the wave form, and its prismatic coefficient and position of center of buoyancy will be as indicated in Table VII. Curves having other prismatic coefficients may readily be constructed by using various other ratios of diameter of generating circle to base.

TABLE VII
AREA CURVE FACTORS

	1	2	3	4
P.C.	.507	.519	.530	.557
C.B.	52.8%	53.4%	53.9%	55.3%
1	.080	.080	.080	.080
2	.285	.285	.285	.285
3	.560	.560	.560	.560
4	.808	.808	.808	.808
5	.972	.972	.972	.972
6	.978	.983	.983	.988
7	.800	.823	.850	.900
8	.447	.507	.557	.677
9	.138	.167	.200	.296

Midship section 56% of *L.W.L.* abaft forward point of immersion.

AREA CURVE FOR POWER BOATS

With power boats of the displacement type having speed-length ratios of less than 2.5, the shape of the area curve and the value of *P.C.* have an important influence on the efficiency of propulsion. The most

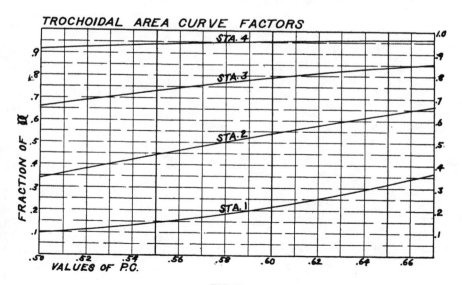

FIG. 28

efficient values of *P.C.* for various speeds are indicated in Fig. 125, and their relation to resistance is discussed in Chapter XV.

The area curve of power boats should have its center of gravity, which is the center of buoyancy of the boat, nearly amidships — from 50 to 53 per cent of *L.W.L.* abaft the forward end. A double trochoid gives a very good distribution of displacement. Fig. 28 gives factors for constructing trochoidal area curves of various prismatic coefficients, half the curve being given with five equally spaced ordinates, as illustrated by Fig. 29 which is a trochoid with *P.C.* = .57. Reading vertically from .57 in Fig. 28 we get factors as follows:

Sta. 1	.17
Sta. 2	.48
Sta. 3	.76
Sta. 4	.95
Sta. 5	1.00

The area of midship section for a given boat multiplied by these factors gives the true area curve for *P.C.* = .57 and areas of intermediate stations may be interpolated.

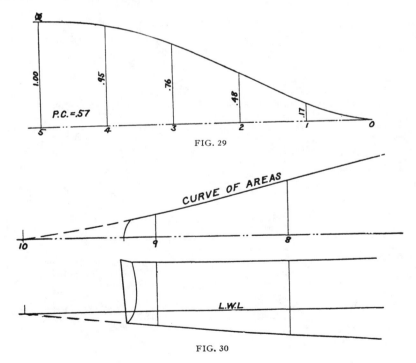

FIG. 29

FIG. 30

Curves of higher prismatic coefficients for afterbody than for fore-body may be used if desired. If necessary to have the *C.B.* abaft midships, the forebody and afterbody will each be divided into five equal intervals, the interval aft necessarily being smaller than the interval forward.

Where speed-length ratio is greater than 2, a flat counter should be used to minimize change of trim and the boat may be designed to have the transom slightly immersed when at rest as shown in Fig. 30. The area curve may be laid out to terminate aft at a point where the water line would end if the hull were extended far enough to avoid any submergence of the transom. This is shown by dotted lines in the figure. The virtual water line to which the boat is designed being thus longer than the actual water line, there results a finer forebody than would otherwise be the case.

From the foregoing the importance of starting a design for a boat of moderate speed by constructing the curve of areas and making all the sectional areas conform to it will be appreciated. Not only is the form for least resistance so obtained but in this way all irregularities in the curve are avoided and the displacement and position of center of buoyancy are sure to work out as intended after the design has been faired up. This method of designing from the area curve is explained more fully in Chapter VIII.

Where extremely high speeds are to be obtained, the shape of the area curve ceases to be of importance.

CHAPTER V

STABILITY

THE statical stability of a vessel is the moment of the couple formed by the weight and buoyancy. In Fig. 31 *WL* is the water line in the erect position, *W'L'*, the water line, when heeled to the angle θ under the action of the wind or some other force, *B*, the position of the center of buoyancy in the erect position, *B'* the position of the center of buoyancy when heeled to the water line *W'L'*; *G* is the position of the center of gravity and *a* is the distance between lines passing through

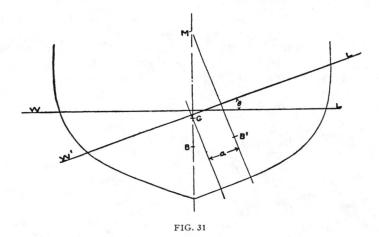

FIG. 31

G and *B'* perpendicular to *W'L'*. The intersection at *M* of the perpendicular through *B'* with the center line is known as the metacenter; *GM* is called the metacentric height, then *a = GM* sine θ. As long as *G* is below *M*, the stability is positive, that is, the vessel will tend to return to the erect position when inclined. With *G* and *M*, the stability is said to be neutral and with *G* above *M*, the stability is negative and the vessel will take an inclination until she reaches a position where *G* is below *M*.

The transverse statical stability of a yacht is equal to the displacement multiplied by *a*, the righting arm. The process of determining the stability at any angle of heel consists in finding the distance apart of the centers of gravity and of buoyancy in a direction parallel to the water line at that angle. It is evident that the stability is dependent upon two factors, that of weight, ballast and construction as affecting the position of G, and that of the form of hull as affecting the position of B'.

The center of gravity, G, lies, of course, on the center line of the yacht in the erect position. Its vertical position on that line may be found by direct calculation, by approximation and by experiment on the completed yacht. The calculation for the vertical position of the center of gravity is similar to that for the longitudinal position as performed in Chapter VI. The weights of the component members of the structure and equipment are found, together with their vertical distances from a reference line, usually the base line or load water line. The moments of the weights are summed up and divided by the total weight, giving the distance of the center of gravity from the reference line. This is a laborious process, too much so for ordinary small yacht work, and it is simpler and generally sufficient to arrive at the position of the center of gravity by an approximation. A simple method of approximating its position is to divide the weight of the yacht into several portions, as, for instance, the weight of the rig taken near the center of effort, the hull proper taken at the center of its profile, the deadwood taken at its center of profile, and the lead keel taken at its center, which may be computed or estimated closely. Adding the moments of these weights about the water line and dividing by the total weight, we have a close approximation to the actual position of the center of gravity. The application of this approximation to the 30-footer is as follows:

ITEM ABOVE L. W. L.	WT.	ARM.	MOM.
Rig......................	1200	21	25200
Hull......................	8980	.8	7180
Equipment................	1800	0	0
Crew.....................	600	3	1800
	12580		34180

ITEM BELOW L. W. L.	WT.	ARM.	MOM.
Deadwood.................	1820	3.0	5460
Lead.....................	9800	4.1	4020
	11620		45660

$$C.G. \text{ below } L. W. L. = \frac{45,660 - 34,180}{12,580 + 11,620} = 0.47 \text{ ft.}$$

After the yacht is built and afloat, the center of gravity may be accurately located by an inclining experiment. Inclining experiments should be performed whenever possible, as the knowledge of the posi-

tion of the center of gravity obtained in this way is of great assistance
in approximating its position for a new design of the same type. Let us
consider, first, the case where an inclination is produced by raising a
quantity of ballast or other weight from the hold to the deck and then
moving it outboard.

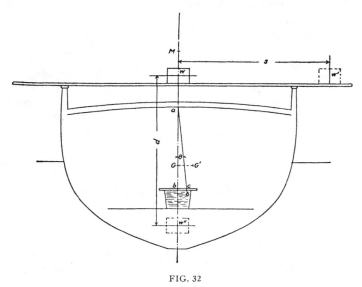

FIG. 32

In Fig. 32, w is the weight which has been raised from the hold
where it occupied the position w''. With the weight at w, the center of
gravity is at some point G. Now if the weight is moved outboard a
distance s, to the position w', the center of gravity moves from G to G',
causing an angle of inclination θ, which is indicated by a plumb bob
attached to a deck beam at a. It is a good plan to have the plumb bob
hang in a bucket of water as shown so as to steady it. The distance it
swings, $b\ c$, is noted on a stick across the top of the bucket.

Now the distance GG', which the center of gravity has moved, is
equal to

$$\frac{w \times s}{Disp.} \text{ for } D \times GG' = w \times s$$

M is the position of the metacenter and is found by computation
from the lines as explained later, then

$$GM = GG' \cot \theta = \frac{w \times s}{Disp.} \times \frac{ab}{bc}$$

ab and bc are readily measured, so that we may solve for GM, giving us at once the height of the center of gravity with the weight on deck at w. If d is the distance w has been raised, the center of gravity with w at w'' will be lowered by the amount

$$\frac{w \times d}{Disp.}$$

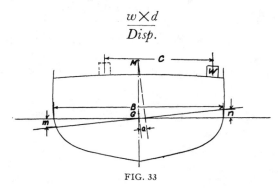

FIG. 33

Where there is no weight aboard which may be used for inclining the yacht, it is necessary to employ some weight which is not a portion of the yacht's equipment. Let M be the position of the metacenter calculated for the immersion with this extra weight aboard. Then as before

$$GM = GG'cot\ \theta = \frac{w \times s}{D+w} \times \frac{ab}{bc}$$

which gives G as the position of the center of gravity with the extra weight on board. Now with the removal of the weight, the center of gravity will be lowered by the amount $\dfrac{w \times h}{Disp.}$ where h is the height of w above G.

Another method involves the moving of a known weight a convenient distance laterally and measuring the amounts of emersion and immersion, m and n, (Fig. 33). Heeling moment $= W \times C =$ righting moment $=$ displacement $\times a$,

$$\frac{a}{GM} = \frac{m+n}{B} \text{ whence } a = \frac{(m+n) \times GM}{B}$$

substituting $\dfrac{disp. \times (m+n) \times GM}{B} = W \times C$

and $GM = \dfrac{W \times C \times B}{(m+n) \times disp.}$

All distances are measured in inches or feet, and W and displacement are in same units, either pounds or tons.

The experiment should be conducted on a still day in a location where there is no sea or current. All bilge water should be pumped out and no one besides the observer should be aboard when the readings are taken. After taking a reading, the weight should be shifted to the other side an equal distance from the center line and a second reading taken. Two different inclinations should be used, taking a double reading for each. The angles should be from one to three degrees.

GM BY CALCULATION

For small angles (up to about 10 degrees) the height of M remains practically constant. The righting arm is equal to GM sin θ and the stability equals $D \times GM$ sin θ.

We have, then, simply to determine the metacentric height in order to find the stability; this is known as the metacentric method of determining stability and is applicable only to very small angles of inclination.

The height of the metacenter above the center of buoyancy is equal to the moment of inertia of the water line plane divided by the volume of the displacement, or

$$BM = \frac{I}{V}$$

As the demonstration of this is somewhat complicated, it is deemed inadvisable to give space to it here, and the reader is referred to works on theoretical naval architecture for this demonstration.

We have seen in Chapter II, page 21, how to determine the moment of inertia of a water line about its longitudinal axis. For the 30-footer, the transverse moment of inertia of the load water line works out at 2025. The height of the transverse metacenter above the center of buoyancy is

$$BM = \frac{I}{V} = \frac{2025}{378.2} = 5.36 \text{ ft.}$$

378.2 is the displacement in cubic feet. Now $GM = BM - BG$, so that we have to locate B and G in order to determine GM. The methods of determining the heights of the centers of buoyancy and of gravity have already been dealt with. In the 30-footer, the center of buoyancy is very nearly one foot below the center of gravity and the righting arm at 10 degrees is then, by the metacentric method, about $4.36 \times .174 = .76$ ft. (GM sin $\theta = a$, .174 being sin $10°$.)

The metacentric height for longitudinal inclinations is analogous to that for transverse inclinations and is found in the same way, with the difference that the moment of inertia of the water line is taken about a transverse axis through its center of gravity. As worked out on page 22, the longitudinal I for the 30-footer is equal to 11,454, and longitudinal BM is then equal to

$$\frac{11,454}{378.2} = 30.29 \text{ ft.}$$

It is evident that for small inclinations the metacentric height is a measure of the stability. This leads to a consideration of suitable values of GM for various classes of yachts. In racing yachts, where the sail area is unlimited, the metacentric height is made as great as possible either by placing the center of gravity very low or by greatly increasing the moment of inertia of the water line. A large metacentric height represents great stiffness and a tendency to return quickly to the upright from small inclinations. This causes violence of motion when among waves and is to be avoided in power-driven craft where the steadying effect of sail is absent. It is a popular error that great stiffness is essential to a comfortable power craft. On the contrary, boats having small metacentric heights are often the most steady at sea. This holds in a way in the case of sailing yachts, and for this reason cruising yachts have in general smaller metacentric heights than racing boats. The value of GM for steam yachts about 100 feet water line should be about 1.5 feet, for steamers of 200 feet water line it lies in the neighborhood of 2 feet.

As stated, the amount of metacentric height is an indication of the stability for very small angles only, and for this reason cannot be used for determining sail-carrying power and is no indication of the range of stability. It is especially useful in investigating the condition of steamers in various states, such as without coal in the bunkers, with coal and stores aboard, and in the launching condition. To have a complete knowledge of the stability of a yacht, we must determine the lengths of righting arm at various and wide angles of heel. If we plot these values as ordinates, using angles of heel as abscissæ, we have a curve of righting arms. This curve for the 30-footer is shown in Fig. 34. From this curve the length of righting arm for any given angle of heel is readily interpolated. The actual stability or righting moment is, of course, obtained by multiplying this righting arm by the displacement. The points to be noted in a stability curve are the angle at which the stability is a maximum, the amount of stability at that angle and the range of stability. The curve of the 30-footer shows a range of 127 degrees, or greater than could possibly occur from effects of wind pressure alone. Many heavily-ballasted keel racing boats have a range of stability of 180°, the stability at no time being negative. On the other hand, lightly-ballasted centerboard boats, while having large metacentric height and great stability at small angles, have a limited range, reaching their limit sometimes at as low as 50° or 60°.

The amount of stability at small angles, as obtained from the metacentric height, is dependent solely upon the position of the center of gravity and the form of the underbody. At large angles, however, the amount of freeboard and general form of the topsides play an important part. An increase of beam increases the height of the first part of the curve of righting arms, while an increase of freeboard increases the length of the curve.

STABILITY METHODS

There are numerous methods of determining stability at large angles of heel, most of which are too cumbersome and laborious in their application to be of value for yacht work. The three methods, a discussion of which follows, have been selected with especial reference to their applicability to small yacht work. These are:

1. *Blom's Method.* This is a purely mechanical method, and consists in pricking off on very thin card from a tracing of the body plan, the shape of the underwater portions of evenly-spaced transverse sections at a given angle of inclination and up to a water line cutting off the required displacement. These are pasted lightly together in their correct relative positions and their common center of gravity determined. This is done by suspending the sections from two or more points at the edge of the card and noting where plumb lines from the points of suspension intersect. This point is the center of buoyancy at that angle, and the distance between the centers of buoyancy and of gravity in a direction parallel to the water line is easily measured. If this process is pursued for other angles at intervals of about 10° we may then draw a curve of righting arms.

2. *Heeled Longitudinal Sections.* This method is by direct computation, and as an example of its application, the righting arm for the 30-footer at 20° heel, has been worked out. The work is recorded in the table on page 55.

In Fig. 35, the water lines, $W'L'$ and $W''L''$ are drawn so as to cut off the same displacement as WL and represent an angle of heel of 20°. Longitudinal sections perpendicular to $W'L'$ and $W''L''$ are drawn at a, b, c, etc. The sections are spaced a foot apart, and section g passes through the center of gravity G, found by the approximation already given. Referring to the table, the measurements recorded in the

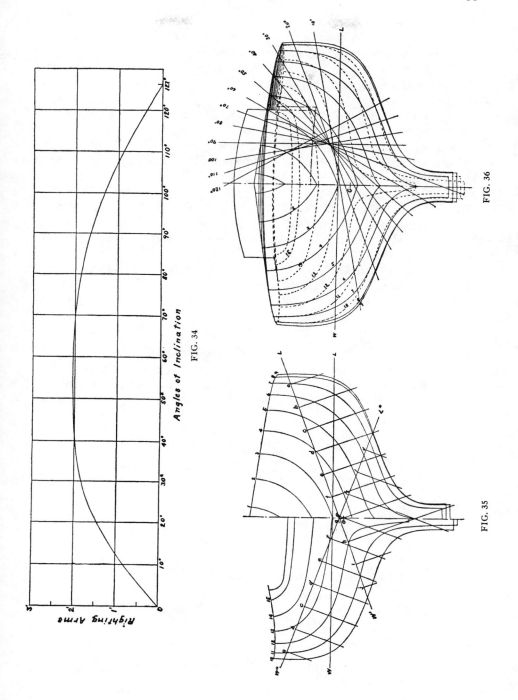

FIG. 34

Angles of Inclination

Righting Arms

FIG. 35

FIG. 36

columns headed a, b, c, etc., are the distances in feet from the inclined water line to the intersection of the longitudinal and transverse sections. By drawing two inclined water lines, $W'L'$ and $W''L''$, we avoid the necessity of drawing a double body plan as in Fig. 36. Measurements for the forebody are taken from $W'L'$ as far as section h, and from $W''L''$ for sections i and j. Measurements for the afterbody are taken from $W''L''$ as far as section i with the exception of station 12 on section h, where the distance from the first to the second intersection by h is measured. Sections i and j are measured from $W'L'$. After the measurements are all taken, the columns are added and their sums multiplied by the spacing of the stations, in this case three feet, give the areas of the sections a, b, c, etc., in square feet by the trapezoidal rule. The areas are then transferred to the column at the right, headed areas, and added together. The sum of the areas multiplied by the spacing of the sections (1 foot) gives for the heeled displacement $374.85 \times 1 \times 64 = 24,000$ lbs.

The area of each section is multiplied by its distance in feet from g, the section passing through the center of gravity G, and the moments on each side of g are then added separately. The difference of the sums of moments divided by the sum of the areas of sections gives 1.41 feet for the length of the righting arm at 20°.

3. *The Mechanical Integrator.* The use of the integrator reduces greatly the amount of labor in making stability calculations, and for this reason is now used practically to the exclusion of numerical methods, by naval architects who have much stability work to do. Fig. 37 is a

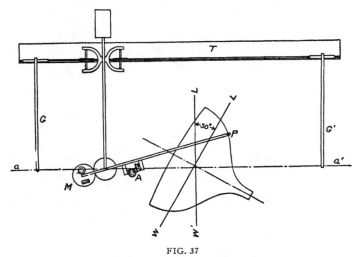

FIG. 37

MEASUREMENTS ON SECTIONS

Sta.	a	b	c	d	e	f	g	h	i	j
4				.06	.39	.50	.49			
5			.52	1.00	1.13	1.10	1.22	.70		
6		.66	1.41	1.73	1.83	1.82	1.94	2.86	.65	
7	.11	1.58	2.07	2.22	2.26	2.27	2.53	4.08	1.26	.37
8	.88	2.13	2.41	2.47	2.48	2.54	2.87	4.26	1.72	.66
9	1.20	2.22	2.47	2.50	2.50	2.55	2.87	4.41	1.72	.67
10	.85	1.92	2.20	2.27	2.22	2.22	2.52	4.16	1.18	.33
11	.10	1.35	1.68	1.72	1.66	1.58	1.96	3.69	.33	
12		.63	1.03	1.06	.90	.74	1.22	.24		
13			.32	.35	.18					
Sum	3.14	10.49	14.11	15.38	15.60	15.32	17.62	24.40	6.86	2.03
Area	9.42	31.47	42.33	46.14	46.80	45.96	52.86	73.20	20.58	6.09

SUMMARY

Sec.	Area	Arm	Mom.
a	9.42	6	56.82
b	31.47	5	157.35
c	42.23	4	169.32
d	46.14	3	138.42
e	46.80	2	93.60
f	45.96	1	45.90
			661.17
g	52.86	0	
h	73.20	1	73.20
i	20.58	2	41.16
j	6.09	3	18.27
Sum	374.85		132.63

Righting Arm at 20° Heel $= \dfrac{661.17 - 132.63}{374.85} = 1.41'$ (S = 1') diagram of a common form of the instrument. The integrator runs on a steel track T and P is the tracing point. The disc at A records area readings, and disc M records moment readings. Some forms of the instrument have still another disc for finding moment of inertia of plane figures.

In using the integrator, the track is first set parallel to and at the correct distance from the axis of the figure by means of the gauges G and G'. In Fig. 37, the instrument is placed for finding the righting arm of a yacht at 30° inclination. The line $a\ a'$ is the axis about which moments are taken. This line passes through the center of gravity G, and makes an angle of 30° with the center line. Although it is convenient to have the axis pass through the center of gravity, it is not essential, for a correction in length of righting arm is readily made, as shown in Fig. 38. Suppose the righting arm about an axis $A\ C$ is found to be $C\ B$. Now, if the center of gravity is at G, the true righting arm will be

$$GB = CB + GC$$
$$= CB + GA \ sin \ \theta$$

that is, if the center of gravity is below the intersection of the axis with the center line, we must add to the arm as found the distance from G to intersection with the axis, times the sine of the angle of inclination. If G is above the intersection, $G\ A\ sin\ \theta$ is to be subtracted.

If we set the discs of the integrator at 0, and then sweep the tracing point around a transverse section, we may take readings of the discs which, when multiplied by the constants of the instrument, give us the area and moment of the section. The moment divided by the area gives the distance from the axis to the center of the section. If this is done for all the sections of the boat and the areas and moments of the sections summed up, we may get the displacement and righting arm. The application of this method requires a double body plan such as is shown in Fig. 36. In that figure the afterbody is drawn in dotted to avoid confusion.

The area and moment readings may be taken and recorded separately, but an easier and quicker way is to sweep the tracing point around each of the sections in succession, without stopping to take readings until all have been traced. Care must be taken not to omit any sections. The final readings give the sum of the areas and moments of the sections. These readings have simply to be corrected for the constant of the instrument and for scale before finding displacement and arm.

The sums of the moment and area readings for the 30-footer at 20°

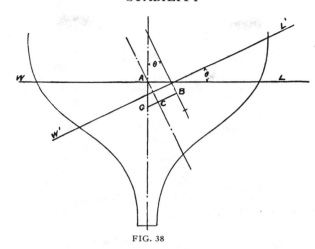

FIG. 38

inclination were respectively 19.00 and 34.99. The constants of the instrument were 4 for moments and 2 for areas. The righting arm is equal to

$$\frac{19.00 \times 4 \times 64/27}{34.99 \times 2 \times 16/9} = 1.45 \text{ ft.,}$$

64/27 corrects the moment reading for scale, being the cube of 4/3, the inverted scale. The area reading is corrected by multiplying by 16/9, the square of the inverted scale.

If the moment disc goes backward, the final reading has to be subtracted from the initial reading, and the center of the figure is on the side of the axis away from the track, otherwise it is on the same side as the track.

The displacement in the inclined position is, of course, the same as when erect. An inclined water line is drawn which it is estimated will cut off the right displacement, and the displacement to this line is calculated. The displacement will be in error by the amount of a slice whose thickness is equal to the amount of the excess or deficiency in displacement to the heeled water line divided by the area of the water line, which is found by measuring the widths of the water line on each station and applying the trapezoidal rule. With this correction, a new water line is drawn parallel to the trial line, which will cut off nearly the right displacement, and the stability for the portion below this new line worked out.

It is sufficient, in general, to determine the stability for a constant displacement as the changes in flotation due to consumption of stores, etc., are slight, especially in sailing yachts. In large, ocean-going

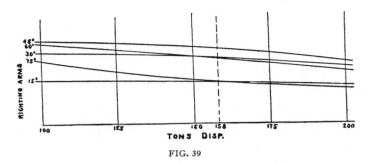

FIG. 39

steam yachts, however, considerable changes in displacement may occur on a voyage due to consumption of stores and fuel, and in such cases it is advisable to work out the stability at various displacements as well as at various angles of inclination.

If we find the righting arm for several displacements at the same angle of inclination, we may draw what is known as a cross curve of stability, having displacements for abscissæ and righting arms for ordinates. This is the converse of the ordinary curve of stability which has angles of inclination for abscissæ, the displacement remaining constant. A set of cross curves is shown in Fig. 39. If we wish to draw an ordinary curve of stability for any given displacement, say 158 tons, we draw an ordinate at that point, and with the values of righting arm for the various angles at the ordinate we draw an ordinary curve, using angles for abscissæ. The cross curves shown in Fig. 39 were drawn with an assumed constant position of center of gravity. The center of gravity alters its position, of course, with a change in loading and displacement, and a correction should be made for this as shown on page 57.

The methods of determining stability which have been discussed assume the yacht to be at rest and in still water. These conditions are, of course, absent when a yacht is under sail, and the actual stability will, in general, be greater than the calculated, due to the support afforded by waves at bow and stern. This difference is great in the case of racing boats with long and low overhangs. For this reason stability work on modern yachts must be considered comparative rather than quantitative; that is, data obtained on sail-carrying power for a given type of boat should be used only for work on boats of that type. For instance, we may wish to know a suitable value for wind pressure, for a particular type, in figuring sail area as shown on page 91. Having found the stability for a yacht of the given type having a suitable sail area, assuming the yacht at rest, we may equate the righting and heeling moments as follows: $Disp. \times arm = A \times h \times p \times cos\ \theta$. A is the sail area, h is the vertical distance between the centers of effort and lateral resistance, and p the pressure of the wind on one square foot of sail. All the quantities in this equation are known except p, which is found readily by solving the equation. In this way, values of p may be found for different types which, used in conjunction with the stability determined in the ordinary manner, enable us to apportion the sail area in accordance with the stability. The stability of steam yachts, when under way, is very nearly as calculated.

In Chapter III, I pointed out that with increase in size of vessel the stability increases much faster than the heeling moment. Stability is the product of displacement and righting arm, and heeling moment is the product of wind pressure on sails and the height of their center above the center of lateral resistance. Since D varies as L^3 and $S.\ A.$ as L^2, it is obvious that stability increases with the size of the boat as the fourth power of a linear dimension, while heeling moment increases only as the cube. It is easy to understand from this why large vessels are apt to be too stiff when loaded even though they are relatively narrow and shoal. Masters of square-rigged vessels sometimes filled the hollow steel yards with water or sand to decrease the stability and increase the moment of inertia, thereby making the vessel easier in a seaway. This is quite different from the case of the small yacht where it is desirable to reduce weight of tophamper as much as possible.

A phenomenon of importance in connection with stability is the fluctuation in apparent or virtual weight of the yacht when among waves of sufficient size to lift her bodily. As the yacht is lifted on a wave, she acquires a certain amount of momentum which tends to decrease the displacement and virtual weight when the wave commences to subside. The virtual weight is a minimum when on a crest, and it is increased correspondingly in the trough. The direct effect of this on the stability is obvious when we remember that stability is the product of weight and righting arm. This variation of stability when among waves, the force of the wind remaining constant, is the chief cause of a yacht's inability to carry the same amount of canvas in rough water as in smooth.

DYNAMICAL STABILITY

So far we have considered only statical stability. We have next to consider dynamical stability. This is the amount of work expended in heeling the yacht to a given angle. This work is done in raising the center of gravity, depressing the center of buoyancy, wave-making, eddy-making, and in overcoming frictional resistance. The last three

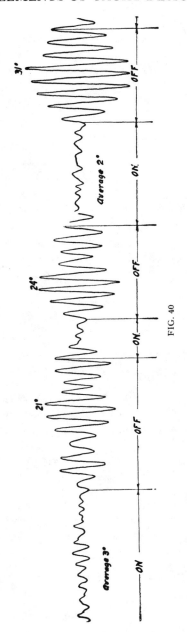

FIG. 40

items increase with rapidity of motion and are of slight importance for slow inclinations. The curve of dynamical stability is the integral of the curve of statical stability; that is, the dynamical stability at a certain angle is equal to the area under the curve of righting moments from zero to that angle and is of importance.

The period of oscillation is the time consumed in swinging from an inclined position to the opposite position under the influence of the stability. It is expressed theoretically by the following formula:

$$T = .554\sqrt{\frac{R^2}{GM}}$$

R is the radius of gyration and GM the metacentric height. It will be seen that T varies as the radius of gyration and inversely as the square root of the metacentric height.

The period of oscillation has an important bearing on the question of speed, as it depends both on the form and the amount and the distribution of the weight, as will be seen from the formula. It appears to be of advantage to make the radius of gyration as large as practicable for transverse inclination, and as small as possible for longitudinal inclinations. This is accomplished by spreading the weights transversely and concentrating them amidships longitudinally, thus increasing the radius of gyration and the value of T. This is the reason for "winging out" the ballast in shoal, centerboard boats. The radius of gyration is increased without materially affecting the metacentric height. Transverse distribution of ballast is not possible in boats having outside ballast.

ANTI-ROLLING DEVICES

One of the chief sources of discomfort in power boats at sea is rolling and there is every incentive for the designer of cruising power boats to turn out a boat which will be as free as possible from this defect. The form and distribution of weights have an important influence on this matter, as the metacentric height depends on these factors. As stated, metacentric heights must be of moderate amount since excessive height causes quick and violent rolling.

There are various anti-rolling devices, among which are the following:

Bilge Keels — These are effective in reducing the rolling angle but offer some disadvantages. They must be placed exactly in the stream line or they will cause excessive resistance. At best, they add considerably to the frictional resistance and this drawback is always present, whether the vessel is rolling or not.

Anti-Rolling Tanks — Various forms of anti-rolling tanks have been extensively experimented with. In them a free fluid is transferred from a tank on the high side to one on the low side at such a time as to tend to check the roll. This scheme does not amount to very much in practice and certainly comes nowhere near stopping the rolling. Besides, there is a good deal of noise from the moving fluid.

Gyroscopic Stabilizer — The Sperry gyroscopic stabilizer is a highly developed scientific device which exerts positive, powerful retarding influence on the rolling of a vessel. A gyroscope of proper weight is rotated at high speed by an electric motor, the axis of the gyroscope being shifted automatically in the fore and aft plane through the controlling action of another smaller gyroscope. This action commences at the very beginning of a roll and kills it before it has developed any considerable amount of energy. The remarkable extent to which rolling is reduced is well shown in Fig. 40, which shows angles of roll of the yacht *Aramis* with the gyroscope alternately in and out of action. The stabilizer fitted to the *Aramis* is shown in Fig. 41, from

FIG. 41

which a good idea of the general appearance of this equipment may be had. The total weight of this stabilizing equipment is usually less than 1 per cent of the weight of the vessel.

There are other decided advantages in eliminating rolling in addition to the increased comfort of those aboard the vessel. These include greatly improved seaworthiness, greatly improved steadiness in steering and increased speed. It is remarkable to note when on a vessel equipped with a stabilizer in a heavy sea, how much drier the vessel is when the stabilizer is put into action and how much steadier a course can be steered. The loss of power from erratic steering is an appreciable one and well worth saving.

Hydroplane Stabilizers — Another method of preventing rolling utilizes stabilizing bilge planes. As shown in Fig. 42, there are two planes installed amidships, one on either bilge, each shaped like a balanced rudder. These are coupled together so as to swing through equal angles. Their action is similar to that of ailerons on airplanes,

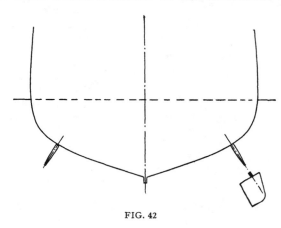

FIG. 42

which are needed to maintain proper lateral angle in flight. These planes naturally exert a powerful heeling effect on the vessel in motion and furthermore are operative in checking the slightest roll or even in keeping the vessel on an even keel when she would otherwise take a list under the pressure of a strong wind. It is true that they add somewhat to resistance when in use but probably less than effective bilge keels do. When not in use, that is when they are turned so that their planes coincide with the stream line, their resistance is negligible. Being balanced, there is little effort needed to incline them, and in small craft they are actuated by a hand wheel similar to a steering wheel. On larger vessels they can be actuated by a hydraulic steerer or other conventional steering mechanism and manually controlled from a wheel, or automatically by a small gyroscope. They are installed amidships, preferably in the engine room.

DIVISION OF STABILITY OF FORM AND WEIGHT

As we have seen, the total stability is made up of the stability due to form and that due to weight. Now, for any given ratio of ballast to displacement, there is probably a certain combination of stability of form and weight more effective for speed under ordinary conditions than any other. This is one of the fundamental principles of yacht design and is, moreover, one of the least understood. The proportion of stability due to weight to the total varies in the same direction as the ratio of ballast to displacement. There is absolutely no data as to what this proportion should be for varying ratios of ballast to displacement. Scientific experiment along this line would be most difficult, if not impossible, but the results would be of immense value to a skillful designer. A faulty distribution of stability between form and weight is responsible for the failure of many racing yachts, the error being generally in giving too large a proportion of stability to weight, that is, in placing the ballast too low.

MOMENT TO TRIM

A knowledge of the longitudinal and transverse metacentric heights is of value in determining the change of trim consequent upon moving a weight from one portion of the yacht to another. The moment to change trim one inch is equal to

$$\frac{D \times GM}{12 \times L} \text{ foot pounds or foot tons.}$$

D is the amount of the displacement in pounds or tons, GM the metacentric height or the distance from the center of gravity to the longitudinal metacenter, and L is the water line length in feet. The change of trim is taken as the sum of the amounts by which the boat sinks at the stern and rises at the bow, or vice versa. The longitudinal GM for the 30-footer is 29.29 ft. The moment to change trim is, then,

$$\frac{24200 \times 29.29}{12 \times 30} = 1969 \text{ ft. pounds.}$$

Suppose we wish to determine the change in trim due to moving a 100 lb. anchor from the bows to the lazarette, a distance of about 29.5 feet. The moment of the weight of the anchor is then $100 \times 29.5 = 2950$ ft. lbs., and the change of trim is

$$\frac{2950}{1969} = 1.5 \text{ inches.}$$

The longitudinal metacentric height is approximately the same as the water line length, so that for rough calculations the moment to change trim one inch may be taken as one-twelfth the displacement in pounds or tons.

MOMENT TO HEEL

The moment to produce one degree heel is equal to $.01745 \times D \times GM$, where GM is the transverse metacentric height and D the displacement, $.01745$ being tan $1°$. For the 30-footer this amount is $.01745 \times 24200 \times 4.36 = 1841$ ft. lbs.

A boat weighing 150 lbs. would be about 8 feet outboard if carried on davits and would produce a heeling moment of $150 \times 8 = 1200$ ft. lbs. The angle of heel produced would be

$$\frac{1200}{1841} = .65°.$$

CHAPTER VI

BALLAST

THE function of ballast is to increase the stability due to weight. This is done by increasing the displacement and by increasing the metacentric height, GM, through a lowering of G. The proper ballasting of a yacht is of vital importance. With cruising boats, considerations of safety and ability demand a liberal amount of ballast. Where speed alone is the only consideration, however, it is desirable to reduce the amount of ballast to a minimum, the sail-carrying power being furnished by increased stability of form and by the weight of crew.

There is little opportunity in this direction, however, under modern rating rules when the displacement is practically fixed and extremes of form are so taxed as to be impracticable.

The nice adjustment of stability of form and of weight, according to type as discussed in the chapter on stability, is of great importance in all cases.

DISTRIBUTION

Lead is more commonly used for ballast than any other material on account of its high density. A secondary advantage is the ease with which it may be cast, bored, smoothed and altered in an ordinary boat yard. Its use permits great concentration and lowering of weight and consequent increase in weight stability. Excess in this direction, however, is a common error. Lead has the further advantage over iron of being uninjured by the action of water. Iron, on the other hand, is much cheaper than lead and may be incorporated in the arrangement for longitudinal strength. Certain types of shoal draft boats, requiring a considerable amount of inside ballast, perform better with stone or iron ballast than with lead. An explanation of this is that the stone and iron, being of smaller density than lead, are of greater bulk for equal weight, thus raising the center of gravity and distributing the weight so as to afford a better combination of stability of form and of weight, and a more suitable value for transverse time of oscillation than would result if lead were used. Transverse distribution and longitudinal concentration of ballast are to be sought in general.

AMOUNT

A convenient method of comparison of the weight of ballast carried in different types is by the ratio of ballast to displacement. This ratio for keel cruising boats should lie between .35 and .50. For the semi-keel type the ratio should lie between .30 and .45, being generally in the vicinity of .40. Centerboard cruising boats with inside ballast should have a ratio of from .20 to .40, the ratio lying most commonly between .30 and .35. It is impossible to state any values of this ratio for racing boats, as the amount of ballast is dependent upon the rule under which the boat is built, and varies from zero in the classes with little or no restrictions up to as high as .80 in classes where the restrictions are more stringent. The ratio ballast to displacement for the 30-footer is .405.

Many people have an idea that a seagoing sailing yacht should carry considerable of her ballast inside to insure easy motion in rough water. This advantage is apparent only in calm weather, and such a boat should be designed for best performance in hard winds. It is important to be able to carry sail above all things and there is little danger of getting a boat too stiff with the average displacement, draft and sail area indicated in Figs. 23, 24, 25, 53, 54 and 60. In the case of large sailing yachts, where it is feared that all outside ballast would produce too great stiffness, most of the weight may be hung outside on the keel and the rest bolted under the deck as far outboard as possible, thus raising the center of gravity and increasing the transverse moment of inertia.

Inside ballast makes a boat cheaper to build but that is about its only advantage. It takes up space, promotes decay and, under some circumstances, strains the hull. When a yacht goes aground, it is obvious that the hull is subject to much less stress if all the ballast is outside. There are many instances of large keel cruising yachts with all ballast on the keel grounding on a ledge and pounding for hours, and getting off later without damage. In like circumstances, a fisherman, with ballast inside, would have lost her shoe, started leaks and possibly become a total loss.

TRIMMING

In all cases it is well to provide for carrying a small portion of the ballast inside the boat. Boats carrying all ballast inside are found to require considerable experimenting with the position and distribution of ballast before securing the best results. From this it seems altogether likely that many racing boats with all outside ballast are not sailing in their best form because the ballast is not in the most suitable position longitudinally or vertically, particularly the former. It is difficult to experiment with the vertical position of the ballast, but the longitudinal position and resultant trim is readily varied if some inside ballast be carried.

An alternate scheme for racing yachts is to have the lead cast in the first place with several detachable chunks along the upper edge, moulded to fit smoothly in place. These are easily removed if necessary to alter trim or flotation, the space left being plugged with wood and smoothed. The lead keel in such case may be designed with a little excess weight.

LOCATION OF C. G. L.

In designs where the entire amount of ballast is to be carried on the keel, it is necessary to make careful computations for weight and position of center of gravity if accurate results are to be expected. To do this, the weight and center of each item entering into the structure and equipment of the yacht must be ascertained. The methods of computing the weights of the various members have been given in Chapter II. Care must be taken that nothing be omitted from this calculation.

The designer should also bear in mind that a boat is constantly increasing in weight throughout her life. This comes from soakage, accumulation of paint and varnish, added fittings here and there, accumulation of supplies, etc., and may amount to as much as 5 per cent in a year or two.

Longitudinal moments must be taken about a convenient axis, usually the midship section, so that the position of the center of gravity may be calculated. The distance from the center of gravity of each member to the midship section, multiplied by its weight, gives the moment. The moments of weights forward of the midship section are summed up separately from those aft and the difference between the sums of moments, divided by the sum of the weights, gives the distance of the center of gravity of the yacht from the midship section. Now, that the boat may trim as designed, the moments of boat and ballast taken about the center of buoyancy must be exactly equal, their weights being on opposite sides of the center of buoyancy. The amount of the ballast is made equal to the displacement minus the sum of the other weights. To illustrate this process the computation for weight and center of gravity of the 30-footer is given on the following pages.

WEIGHTS FORWARD OF STATION 9

ITEM	WT.	ARM	MOMENT
Keel	525	7.5	3940
Planking	2182	1.15	2515
Frames	1104	1.15	1270
Deck	1130	.03	34
Deck Beams	277	.03	8
Shelves	114	2.	228

Clamps	133	2.3	306
Stringers	179	2.	358
Floors	350	5.	1750
Plank Fastenings	95	1.15	109
Deck Fastenings	62	.03	2
Other Fastenings	190	.10	19
House Deck	135	1.	135
House Coaming	223	1.8	402
House Beams	67	1.	67
Mast Step	52	9.5	493
Deck Straps	35	10.4	364
Rail	49	2.3	113
Knees	75	1.	75
Breasthook	18	23.	413
Centerboard	170	1.8	306
Bitt	14	19.6	274
Paint	150	1.	150
Cabin Floor	136	2.4	327
Cabin Floor Beams	18	2.4	43
Deck Fittings	50	2.	100
Mast	517	9.5	4910
Bowsprit	46	24.6	1130
Staysail	21	14.3	300
Jib	23	20.	460
Blocks	55	8.	440
Running Rigging	85	9.	765
Standing Rigging	90	11.5	1035
Windlass	50	19.5	973
Anchors	160	19.	3040
Cables	200	18.	3600
Lockers	270	.04	11
Ice Box and Ice	275	9.2	2530
Washbowl	25	4.	100
Water Closet	90	7.7	693
Tanks and Water	800	2.7	2160
Galley Fittings	80	6.	480
Pipe Berths	30	13.3	399
Bulkheads	232	1.6	371
Personal Effects	250	1.5	375
	10832		37573

WEIGHTS ABAFT STATION 9

ITEM	WT.	ARM	MOMENT
Deadwood	1820	1.7	3095
Horn Timber	61	14.5	885
Sills	168	2.4	403
Transom	41	19.8	812
Tie Rods	40	.2	8
Rudder Tube	10	11.8	118

Cockpit Coaming	170	10.6	1800
Steering Gear	30	12.2	366
Mainsail	116	4.	464
Main Boom	171	8.3	1420
Gaff	66	1.	66
Carpet	35	1.2	42
Berths	225	2.1	472
Cushions	15	2.	30
People	600	10.	6000
	3568		15981

Now dividing the difference of the moments by the sum of the weights

$$\frac{37{,}573-15{,}981}{10{,}832+3568}=\frac{21{,}592}{14{,}400}=1.50$$

we have 1.50 feet for the distance the center of gravity is forward of the midship section. Now the center of buoyancy is 1.58 feet forward of the midship section so that the center of gravity comes .08 feet abaft the center of buoyancy. The weight of the lead will be made equal to the displacement minus the sum of the weights, $24{,}207-14{,}400=9807$, or about 9800 lbs. It would be desirable to carry a portion of the ballast inside for trimming, but for purposes of illustration it has all been placed on the keel in this design. The moment of the ballast must, of course, equal the moment of all the other weights, then

$$9800\times arm=14{,}400\times.08$$
$$arm=\frac{14{,}400\times.08}{9800}=.12 \text{ ft.}$$

The center of gravity of the lead must then be .12 feet forward of the center of buoyancy or .20 feet forward of station 8½, so that the boat may trim properly.

CALCULATION OF LEAD

The next step is to determine the top of the lead on the design so that it will have the required weight and position. To do this a trial line is drawn and the weight and center of gravity of the portion thus cut off is determined. With this as a guide, another line is drawn which it is estimated will correct the errors in weight and center of the first trial line. If there is still an error, another trial is made, and so on until the correct line is found. The volume of the lead is found in the same manner as the volume of the hull in the calculation for displacement. Keel sections are drawn at frequent intervals, 1.5 feet in Fig. 57, and the areas of these sections are taken with a planimeter. The volume and center are then obtained by the application of the trapezoidal rule. Where there is a slot cut through the keel for a centerboard, the volume of the portion cut out is determined and its moment is subtracted from the total moment. The computation for the 30-footer is ås follows:

STA.	PLAN. READING	ARM.	MOMENT
6	.01	5	.05
6½	.17	4	.68
7	.35	3	1.05
7½	.42	2	.84
8	.44	1	.44
			3.06
8½	.40	0	.00
9	.36	1	.36
9½	.32	2	.64
10	.25	3	.75
10½	.19	4	.76
	2.91		2.51

Total weight $= 2.91 \times 2 \times 16/9 \times 1.5 \times 710 = 11{,}020.$

The multiplier 2 is for both sides, 16/9 is the square of the scale, inverted, 1.5 is the common interval and 710 is the weight in pounds of one cubic foot of lead.

$$\text{Center of gravity} = \frac{3.06 - 2.51}{2.91} \times 1.5 = .283 \text{ ft. for'd sta. } 8\tfrac{1}{2}.$$

Now the area of the portion cut out by the slot is 5.72 square inches on the drawing, or $5.72 \times 16/9 = 10.17$ sq. ft.

The thickness is two inches or .167 feet. The weight cut out is then $10.17 \times .167 \times 710 = 1202$ pounds. Its center of profile is .95 feet forward of station 8½. Now, taking moments and subtracting,

$$\begin{array}{r} 11020 \times .283 = 3119 \\ 1202 \times \ .95 = 1142 \\ \hline 9818 \times \ .20 = 1977 \end{array}$$

we get .20 feet for the distance which the center of gravity of the lead is forward of station 8½.

It is well to draw the lead on a larger scale than the lines with half sections at frequent intervals as shown in Fig. 43 and plot a curve of lead areas as shown in Fig. 44. Greater accuracy will be had than if the work is all done directly from the drawing of the lines.

The calculation for the longitudinal position of the center of gravity is necessary when all the ballast is to be carried on the keel, to assure satisfactory results. The relative positions of the center of buoyancy

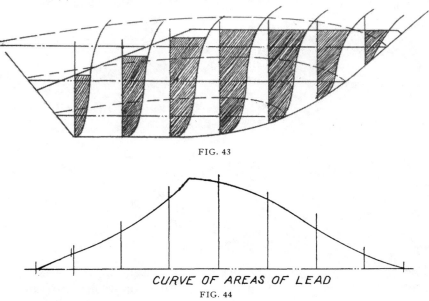

FIG. 43

CURVE OF AREAS OF LEAD
FIG. 44

and of gravity vary so widely in different designs that an estimate of the proper position for the center of he ballast is apt to lead to disappointment. The many cases of faulty trim which are constantly occurring may be attributed to carelessness in this matter. It is a gratification to the designer when a boat after launching and being completely equipped floats exactly as designed. This should always be the case if the designer's work and inspection have been carefully done and the builder has done his part in faithfully executing the plans.

CHAPTER VII

THE LATERAL PLANE OF SAILING YACHTS

THE term lateral plane is applied to the vertical longitudinal projection of the underwater body of the vessel. The center of lateral resistance is the point at which the lateral pressure of the water on a boat sailing close hauled may be considered to be concentrated. In reality it is practically impossible of determination and is considerably farther forward than the center of figure of the lateral plane. The reasons for this are that as the boat moves forward she is constantly entering solid water, increasing the pressure at the bow and decreasing it toward the stern, where the water is more or less disturbed. The wave which is piled up under the lee bow also tends to increase the pressure in the region of the bow. Moreover, the contour of the lateral plane when the yacht is heeled is quite different from that in the erect position. The position of the center of buoyancy is also considered to affect the position of the center of lateral resistance.

CENTER OF LATERAL PLANE

Since we cannot determine the true center of lateral resistance we must assume a center for reference in placing the center of our sail plan. The center of lateral plane is generally used for this purpose, although some designers use a center lying on a line between the center of buoyancy and of the fin or centerboard, entirely disregarding the hull in shoal-bodied boats. This point probably bears a more constant relation to the true center of lateral resistance than does the center of lateral plane. In either case, the proper fore and aft distance from the reference point to the center of the sail plan must be determined empirically from existing boats of the given type.

It is found that to secure satisfactory results the fore and aft positions of the center of the lateral plane must lie within certain definite limits. Its position is commonly located with reference to the forward point of immersion, its distance from that point being expressed in terms of the water line length. In most yachts, this distance is between 54 and 59 per cent of the water line. In the majority of cases, 56 or 57 per cent will give the best results. In cases where the center of the sail plan is necessarily very far aft, as in the old-fashioned catboats, the center of lateral plane may be placed as far back as 60 or 61 per cent. In centerboard boats, it is well to have the center of the board somewhat ahead of the general center so that when the board is raised the general center may move aft, making the boat steer better off the wind.

The necessary amount of lateral plane is something for which no definite rules can be stated, as it varies greatly with the type of boat; thus a centerboard boat requires relatively less lateral plane than a keel boat, as the centerboard, being a plane surface, is more effective in resisting lateral pressure than the somewhat rounded surface of the keel. The type of rig and amount of sail carried also have a bearing on this question. A convenient comparative method of approximating the area of lateral plane is by the ratio of sail area to area of lateral plane. This ratio, worked out for a large number of representative yachts, is found to vary from 6 to 7. The lower ratios are found in heavy cruising boats. For keel boats of ordinary type, the ratio should lie between 6.25 and 7.0. Centerboard boats may, in general, have a slightly larger ratio than keel boats of similar type. Another method is to apportion the area of lateral plane by its ratio to the area of the midship section. This ratio may vary from 4.0 to 6.0. For a keel yacht, it may be from 4.25 to 5.50. The seagoing yacht requires an area considerably in excess of the amount necessary for "holding on" in smooth water.

Bilge boards are much more effective in resisting lateral pressure than a centerboard as they are nearly vertical when the boat is heeled down when beating to windward and present a surface normal to the direction of pressure. On this account the area of each bilge board may be made much less than the necessary area of a centerboard for the same boat (about .7 as great). The decreased wetted surface and capsizing tendency of bilge boards indicate their use on shoal racing boats.

Designers differ as to whether the rudder should be considered a part of the lateral plane or not. In a properly balanced yacht, there is some pressure on the rudder or weather helm, when sailing by the wind, and for this reason the rudder may be considered a factor in resisting lateral pressure. This is especially true in some types, such as catboats, where it is impracticable to secure proper balance by placing the rig far enough forward. The angle of weather helm and consequent loss of speed in such boats is reduced by increasing the area of the rudder so that the rudder is truly lateral plane in boats of this type.

The area and contour of the lateral plane have an important influence on the performance of the yacht. In the racing yacht, especially, the area and distribution of plane must be adjusted to a nicety. It is obvious that excessive area adds to resistance because of increased friction, although its influence is less than one would think. An instance of this is the case of two yachts in a closely restricted class—very evenly matched in good winds and of practically the same sail area. The one which had slightly greater displacement, length and sub-surface area was, nevertheless, distinctly faster in very light airs. From this we infer that economy in wetted surface at the expense of easy lines, holding-on power, proper balance or other qualities does not pay.

CENTER OF LATERAL PLANE MOVES AROUND

Lateral resistance is a complex matter inasmuch as a large fraction of it is contributed by the rounded surface of the hull itself and the

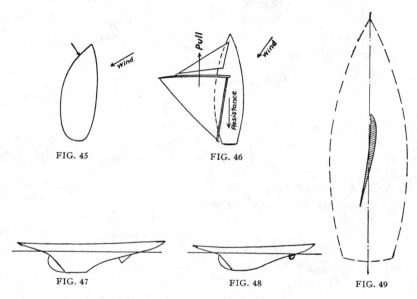

FIG. 45   FIG. 46

FIG. 47   FIG. 48   FIG. 49

amount and center of application of hull resistance varies with the angle of heel. The shifting of the center varies also with the type of boat pretty much as the character of water line varies. With a beamy, shoal draft craft such as the Cape Cod catboat, where the inclined water line is quite unsymmetrical, as in Fig. 45, the center shifts forward greatly when heeled, causing the boat to carry a strong weather helm. The old-fashioned narrow English cutters had an opposite tendency.

In the case of keel yachts, at least, it is possible to form the hull so that the center of lateral pressure does not shift with angle of heel, or even so that it shifts aft slightly to counteract the rounding-to couple of the horizontal fore and aft component of the wind pressure and the resistance of the boat to forward motion. With high rigs, the fore and aft component of the wind pressure is quite far outboard when the boat is heeled; this, with the hull resistance, creates a couple tending to bring the boat into the wind. This is illustrated by Fig. 46, which is a view from above of a yacht on the starboard tack heeled well over.

It is clear that in light winds, when the boat is nearly on an even keel, this turning moment is reduced to little or nothing, hence the principal difficulty in designing a boat to balance perfectly in both light and strong winds. The natural course is to balance for slight lee helm in extremely light airs. A good expedient in small racing boats is to arrange the mast to be raked aft easily when wind is light, thus throwing the center of effort aft. (See Fig. 110.) Another way of reducing the fore and aft distance between centers of effort and lateral resistance is to shift the latter center forward in some manner. In small craft the centerboard is sometimes arranged to shift forward, thus correcting any tendency toward lee helm.

FORWARD CENTERBOARDS AND RUDDERS

In isolated instances, this has been accomplished in larger craft by a small centerboard fitted forward, as shown in Fig. 47. This board would naturally be used only in light winds for correction of lee helm or when hove to in heavy weather.

Suppose this small bit of forward lateral plane were made so that it could be set at an angle with the keel. This would, of course, increase the pressure on it and consequently its effectiveness. For correcting lee helm, it would naturally be set with its after edge to leeward. As the wind strengthens and the boat gradually acquires a weather helm, this plane, if adjustable, would have its after edge swung to windward. By reducing this plane in size, because of its increased effectiveness due to variable angularity, it logically develops into a bow rudder, as shown in Fig. 48. This, to my knowledge, has never been used on sailing yachts. That, of course, is no reason why it should not be, and I offer the idea as one containing possibilities for certain types of sailing yachts. A bow rudder could be directly connected with the regular rudder to swing equal angles, but in the opposite direction, or it could be controlled separately and used as an adjustment to correct lee or weather helm as these varied with different strengths of wind. Being far forward and in a position exposing it to such direct pressure, the bow rudder is extremely effective and the probability is that in cases of bad weather helm the bow rudder, with its great leverage, would greatly decrease the angle of the main rudder, with consequent decrease in resistance. It would also have a great steadying influence in a seaway, and in tacking would enable the boat to tack more quickly and with less loss of headway, due to her turning about a point near the

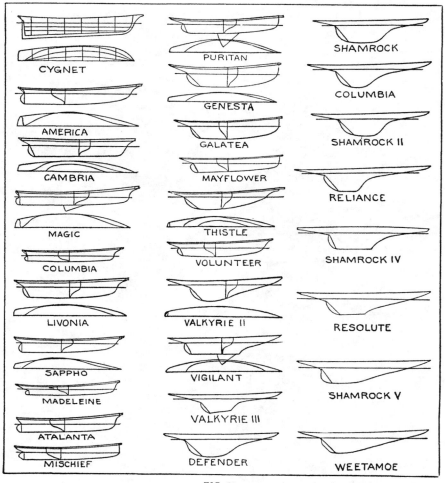

FIG. 50

center of gravity rather than a point much farther forward, as is the case when turning pressure is all exerted aft.

The sloping profile of lateral plane similar to that shown on the *Gypsy* (see Fig. 110), seems to be effective in promoting good balance for all wind strengths. A lateral plane of this sort seems to offer the

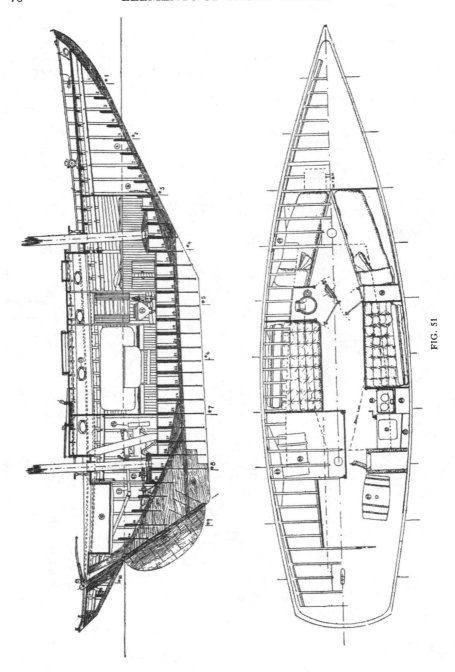

FIG. 51

best distribution of area for effective windward work under all conditions.

The forward portion of the lateral plane is most effective for lateral resistance, and for this reason the leading edge should be made as long as possible. To carry out this idea, the keel bottom should have considerable drag or slope, as then the leading edge extends from the bow aft to the rudder and is constantly entering solid water all along its length.

GREAT CHANGES IN CONTOUR OF LATERAL PLANE

It is interesting to trace the development in contour of lateral plane over a period of ninety years, as illustrated in Fig. 50. A few years back, there was a great tendency to concentrate the lateral plane amidships.

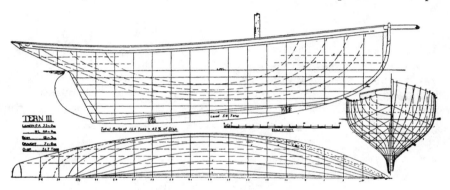

FIG. 52

This, of course, promotes quick turning, but on the other hand makes a boat unsteady in her steering, particularly in a seaway, and the modern tendency is to spread the lateral plane fore and aft a little more, in both racing boats and cruisers. For real seagoing work, the lateral plane should be generous and well-spread out fore and aft. The *Stormy Petrel*, Fig. 51, illustrates a conservative lateral plane for cruising boats. The lateral plane on *Tern III*, shown in Fig. 52, represents a still more conservative English type designed for heavy weather conditions. Such a boat as this will run and steer easily under all conditions and will lie hove to under sail or to a sea anchor.

The fin on keel yachts is necessarily symmetrical transversely, as the boat must sail equally well on either tack and pressure comes alternately on either side. One must realize when he considers other pressure surfaces, such as the aeroplane wing, propeller blade, turbine blade, etc., which always take the pressure on one side, that a blade with identical face and back must be much less efficient than one designed with a special flat or concave pressure face. If a boat were to sail always on one tack, her fin could be designed with a face and back as shown in Fig. 49 for a boat sailing on the starboard tack. Such a fin, set with the proper angle with the longitudinal axis of the boat, would prevent all leeway and offer the least resistance in the direction of motion.

It is obviously impracticable to sail always on one tack and the mechanical difficulty of constructing a fin that can be made to alternate the convexity and concavity of its faces when coming about appears insurmountable. There are, however, possibilities in the way of employing properly designed pressure surfaces in the case of lee boards, bilge boards and twin centerboards. There is much room for profitable experiment along this line.

MEASURING THE AREA

The exact form of the calculation for area and center of lateral plane depends on the character of the design. If the plane is regular in contour, the trapezoidal or Simpson's rule may be used. If such a plane has a centerboard, the areas and centers of hull and board should be found separately and the combined center found by taking moments about a convenient point. In the case of an irregular contour, the plane should be divided into a number of portions so that the area and center of each may be readily figured. Moments of each are then taken about a convenient point, the sum of the moments divided by the sum of the areas giving the distance from the point to the total center. The cardboard method of determining the center of lateral plane is susceptible of sufficient accuracy, and is much more expeditious than calculation. The calculation of area and center of lateral plane of the 30-footer is as follows. The first column gives measurements on each station in inches from the water line to the keel bottom:

STA.	ORD.	RM.	MOM.
4	.60	5	3.00
5	1.27	4	5.08
6	2.40	3	7.20
7	3.35	2	6.70
8	3.50	1	3.50
			25.48
9	3.60	0	0
10	3.73	1	3.73
11	3.85	2	7.70
12	2.36	3	7.08
	24.66		18.51

$$Area = 24.66 \times 4/3 \times 3 = 98.64 \text{ sq. ft.}$$

Distance center is forward of station 9 is

$$\frac{6.97}{24.66} \times 3 = .85 \text{ ft. or } 17.15 \text{ ft. aft of 3.}$$

This area neglects a small triangle at the forefoot between stations 6 and 7, also one at the after end of keel between stations 11 and 12. It is too great by the amount of the triangle whose hypothenuse is a straight line from the water line on 14 to the stern post on 12. These errors are due to the fact that the trapezoidal rule assumes the figure to be straight between ordinates. To make these corrections let us take moments about station 3.

$$98.64 \times 17.15 = 1692$$

1st triangle	$.86 \times 10.50 =$	9

$$2\text{nd triangle} \quad 1.67 \times 25.35 = \quad 42$$
$$\overline{101.17} \qquad\qquad \overline{1743}$$

$$3\text{rd triangle} \quad 1.32 \times 28.64 = \quad 38$$
$$\overline{99.85 \times 17.07 = 1705}$$

The center of lateral plane, exclusive of centerboard, is 17.07 ft. abaft station 3 or 56.9 per cent. The board has an area of 12.80 sq. ft. and its center is 16.70 ft. back. Taking moments of hull and board

$$99.85 \times 17.07 = 1705$$
$$12.80 \times 16.70 = \ 214$$
$$\overline{112.65 \times 17.04 = 1919}$$

The total *C. L. P.* is then 17.04 ft. abaft station 3 or 56.8 per cent.

CHAPTER VIII

DESIGN

BEFORE starting a design it is well to fix upon the principal dimensions, which include length over all, length on the water line, beam on the water line, extreme draft, area of lateral plane, displacement and area of midship section. Having settled upon the type and water line length, the other proportions are generally arrived at by comparison with existing boats.

SAILING YACHTS

As an aid in making such comparisons, the curves in Figs. 53, 54 and 55 have been prepared. These curves are based on data taken from a large number of successful existing yachts and may be considered thoroughly representative of current American practice. Fig. 53 gives average values of beam in feet on the water line for cruising boats, and Fig. 54 gives extreme drafts. For this work, yachts have been divided into three classes, viz., keel boats with all outside ballast, semi-keel boats with a small centerboard and all or nearly all the ballast outside, and centerboard boats having all or nearly all the ballast inside. It will be noted that the differences between the types, so marked in the smaller sizes, disappear almost entirely in the larger boats. The small sizes vary greatly as to beam and draft and the first part of the curves simply represents a fair average value for these quantities. In general, a boat having greater beam than is shown by Fig. 53, will have less draft than is indicated in Fig. 54 for the same water line length, and vice versa. The larger yachts conform more closely to the curves. Fig. 55 gives average least freeboard to top of rail. Figs. 23, 24 and 25 give average values for displacement, being expressed in pounds up to 60 feet water line, and in tons of 2,240 pounds above that size.

These curves will be found of great value in blocking out the proportions of a new design. By their use, comparison is made of the proportions of a large number of existing yachts of the given size instead of with only one or two, as is generally the case. It is understood that these curves apply only to cruising boats. It should not be inferred that the dimensions of a well-proportioned boat must lie on or close to the curves, as many excellent designs show quite a divergence from them, especially in the smaller sizes. On the other hand, if the proportions

82

indicated by the curves for the given water line length are adhered to, a design may be produced which will be of excellent proportions in the light of present practice.

To illustrate the process of laying out a design, the procedure followed in the case of the 30-footer will be given. Having settled upon a 30-foot water line semi-keel cruiser of normal proportions, the prin-

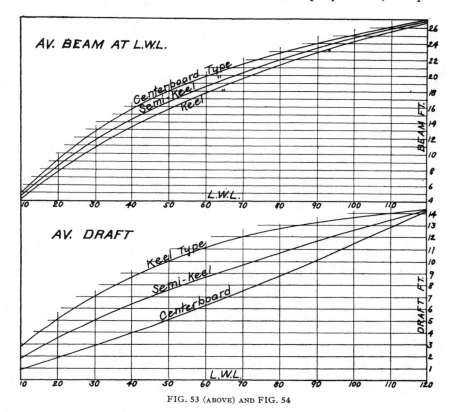

FIG. 53 (ABOVE) AND FIG. 54

cipal dimensions were taken from the curves, the beam at the water line as indicated on Fig. 53 being just under 12 feet. Fig. 54 gives about five feet draft for 30 feet water line length. Fig. 55 gives for the least freeboard just under two and one half feet. The displacement curve (Fig. 24) shows about 23,000 pounds for 30 feet water line; 24,000 pounds was taken as a safe figure. The next step is to determine the proper area of midship section to give this displacement. A suitable prismatic coefficient for this type of boat is .50 to .52. Using .52, the area of midship section should be

$$\frac{\dfrac{24200}{64}}{30 \times .52} = 24.2 \text{ sq. ft.}$$

The half area in square inches on the drawing will be $24.2 \times \frac{1}{2} \times 9/16$ = 6.81 sq. in. for a scale of $\frac{3}{4}$ inch to the foot. The area of lateral plane should be about $24.2 \times 4.65 = 112.5$ sq. ft., as explained on page 74.

We now have sufficient data to enable us to block out the design. It will be assumed that the reader has some knowledge of lines and is familiar with the meaning of the terms elevation, half-breadth plan,

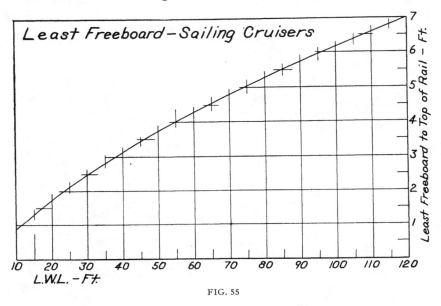

FIG. 55

body plan, etc. The first step is to draw the profile, using the least freeboard, draft, area and center of lateral plane already fixed upon. It is a good plan to draw a preliminary profile at half the scale of the lines and enlarge the final profile from that. The object in doing this is to reduce the drawing to such a size that the eye can take in the entire drawing at a glance and study the way in which each portion harmonizes with the whole. It is well to draw the sail plan first. In this way, the effect of the rig on the general appearance of the boat may be studied and the sheer line, profile of bow and stern, etc., drawn so as to harmonize in the best manner. On the preliminary profile, the freeboard at bow and stern and the length and contour of the overhangs are determined. The contour of lateral plane may also be drawn in on this plan, being altered and redrawn until its area and center come as intended. The final profile may then be enlarged from the preliminary sketch.

The midship section should be drawn in next at the proper distance from the forward point of immersion. This distance was made 56 per cent of the water line length = 16.81 ft. for the 30-footer. The section is drawn with the water line beam, draft and area already assigned and with the height of rail and draft given by the profile at that point. The shape of the section is of great importance, as it influences the form of the entire boat. The most suitable shape of midship section is found by experience. It must possess the elements of speed, stability and sea-worthiness combined in the proportions best adapted to the type of boat in hand. In general, within certain limits, the shoal flat section contains the elements of speed in higher degree than a narrower and more V-shaped section. This statement has numerous exceptions.

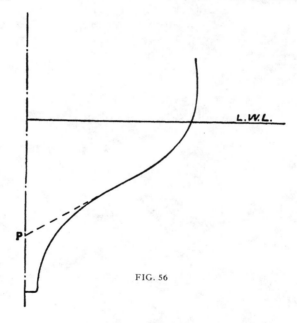

FIG. 56

Having drawn in the midship section, we may next direct our atten-tion to the half-breadth plan. The midship section gives us a point amidships through which to draw the sheer or deck line. The best shape for this line is also a matter of judgment, based on experience. It is a good plan, generally, to have the widest point on deck a little abaft the widest point at the load line, giving what is known as a raking midship section. A preliminary load water line may be drawn next and the half-breadths at deck and load line transferred to the body plan. We now have three points on each station on the body plan through which to draw the transverse sections, *viz.*, the top of plank-sheer, load water line and keel bottom. The usual procedure from this point is to draw in the stations by eye, fair them up, draw a displacement curve and see if the displacement comes out near enough. A much better method is to draw an arbitrary displacement curve to start with and make the area of each station conform to that indicated by the curve for that station. This course was pursued in designing the 30-footer. It is somewhat difficult to preserve the area of the sections during the process of fairing up, but it may be done with care and practice. Where the yacht has an appendage, as is usually the case, we must deal with the areas of sections above the center line. By center line is meant a line passing through the points *P* on each section (Fig. 56), where the prolongation of the flat of the bottom intersects the central plane. The areas above the center line are the areas above the dotted line, all below that line being considered the appendage.

The 30-footer's half area of midship section, in square inches, on the drawing is 5.89. A boat of her type should have a prismatic coefficient for the portion above the center line of about .52. In constructing the curve of areas above the center line, the factors in column 2, table VI were used. This gives for areas of half sections above center line, in square inches:

STA.	½ MID. SEC. SQ. IN.	FACTORS	½ A SQ. IN.
4	5.89	.080	.47
5	"	.285	1.68
6	"	.560	3.30
7	"	.808	4.77
8	"	.972	5.73
9	"	.983	5.79
10	"	.823	4.85
11	"	.507	2.98
12	"	.167	.98
			30.55

Disp. above C.L. $= 30.55 \times 2 \times 16/9 \times 3 \times 64 = 20,900$ lbs.

The stations were drawn and altered until the area of each was as indicated above, and these areas were adhered to in fairing up. The curve of areas above center line is drawn in on Fig. 57 with the curve of areas below center line superposed. This shows markedly the decided bump amidships in the curve caused by the appendage.

Having drawn in and faired the upper body, the appendage may be drawn in and its displacement computed. A slight variation in the displacement of the appendage is easily effected if necessary to bring the total displacement to the desired figure.

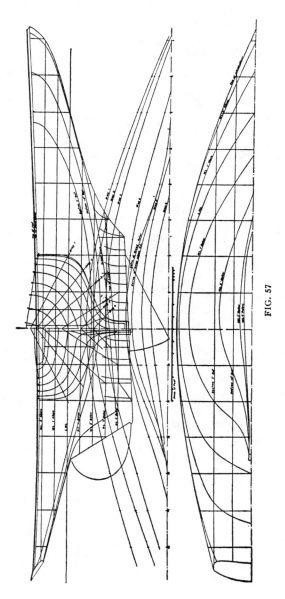

FIG. 57

FAIRING THE "LINES"

A brief outline of the process of fairing lines may be of value at this point. The design is faired by taking sections through the boat in various planes, longitudinally on water lines, buttocks and diagonals, and transversely on the stations. After the transverse sections have been drawn in as nearly fair as possible by eye, the diagonals, buttocks, etc., are developed and the sections are altered until all are perfectly fair curves. The operation of fairing should be commenced with those longitudinal sections whose planes are most nearly perpendicular to the contour of the transverse sections. These are generally the diagonals. Having faired the boat by the diagonals, the water lines or but-

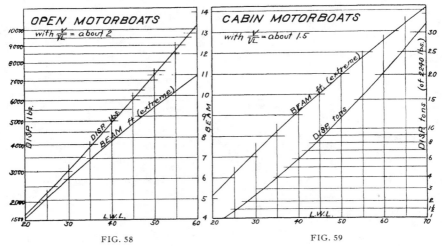

FIG. 58   FIG. 59

tocks are developed next. It is important to take a sufficient number of these, as a boat may be apparently perfectly fair by the diagonals and yet exhibit peculiar places in the ends when the water lines come to be drawn in, and vice versa. The intersections of water lines and buttocks in the elevation must correspond with those in the plan. The procedure is varied somewhat according to the type of boat; thus, in a shoal, flat craft some of the buttocks would be developed first. The designer, as he gains experience, varies the procedure and adopts methods which he has found of especial value to him.

The practice of making a model from the lines is to be recommended, as sometimes slight imperfections in the design are made apparent by the model, where it would be difficult or impossible to detect them in the lines. The practice of making the model first and the lines from the model is now practically obsolete.

After completing the drawing of the lines, the computations for weight, stability, etc., should be made to ascertain whether or not the design will fulfill requirements satisfactorily. These computations are dealt with elsewhere. The final operation is to take offsets from the design from which to lay down the boat full size in the mould loft.

POWER YACHT DESIGN

The process of design as outlined for sailing yachts applies as well to motor yachts, but with them we have no appendage to deal with. Average values of beam and displacement for open motorboats are shown in Fig. 58, and for medium speed cruisers in Fig. 59. Average displacements for various speeds are indicated in Fig. 20, Chapter III, but this is only a guide as to what is usual in the large sized yachts. The proper displacement must be determined to carry the necessary weights of structure, equipment, power plant, fuel and stores, there being little or no ballast for which to provide flotation.

After determining the necessary displacement, the remaining characteristics needed to design the vessel, such as proper prismatic coefficient, area of midship section, shape of area curve, shape of midship section, shape of $L.W.L.$, type of stern, etc., are determined by methods indicated in Chapter XV. A close adherence to the values indicated is necessary for economical performance, and there is much greater incentive for attainment of economical propulsion with power driven craft than with sail, for fuel costs money whereas the wind is free.

CHAPTER IX

THE SAIL PLAN

THE simplest method of determining the amount of sail for a new design is by direct comparison with some boat of known performance and of approximately the same type. This is the usual method. As an aid in estimating the sail area, the curve in Fig. 60 has been drawn,

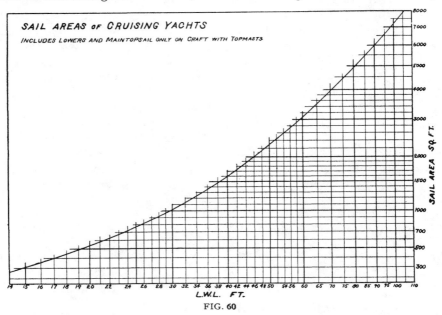

FIG. 60

using water line length as a basis of comparison. The sail areas of a large number of successful yachts (over 300) were used in constructing this curve, and it may be considered thoroughly representative of modern practice for cruising boats. It was found that no distinction could be made between keel and centerboard boats, especially above 20 feet water line. Using this curve for the 30-footer, we get 1000 square feet as a suitable area.

Other methods of comparison for sail area are by the ratio of sail area to area of wetted surface, and of sail area to area of midship section. The former ratio is generally about 3, though it varies between 2 and 4 and increases with size of the boat. The ratio of sail area to area of midship section is used a great deal by French naval architects, and is a good method of comparison. This ratio should be between 45 and 55 for cruising boats. For the 30-footer, it is 53.2. The ratio of sail area to displacement is also used. Theoretically, the sail area should vary as the two-thirds power of the displacement. Let S_1 and D_1 be the sail area and displacement of a certain boat, and S_2 and D_2 the sail area and displacement of another boat of exactly the same form but of different dimensions; then

$$S_1 : S_2 = D_1^{\frac{2}{3}} \cdot D_2^{\frac{2}{3}} \text{ or } S_2 = \frac{S_1 D_2^{\frac{2}{3}}}{D_1^{\frac{2}{3}}}$$

The correct procedure, theoretically, is to apportion the sail area in accordance with the stability, making the heeling and righting moments equal for a reasonable angle of heel, say, 20°. The righting moment is equal to the product of the displacement and the righting arm for the given angle, as explained in Chapter V; while the heeling moment is equal to the pressure on the sails, or pressure per unit area times the area, multiplied by the vertical distance between the centers of effort and lateral resistance. This distance decreases, of course, as the angle of the heel increases, varying as the cosine of the angle. The pressure of the wind on sails for a good whole-sail breeze is generally considered to be a little over one pound per square foot of area, say 1.15 pounds. This is not the absolute pressure of the wind, such as mechanical engineers use in designing structures, but a sort of constant, determined as on page 58. The assumptions made here are that the sails are perfectly flat surfaces, lying in the central plane of the boat, and that the wind blows in a direction perpendicular to the longitudinal axis of the boat. These conditions are, of course, not realized, and for this reason p must be considered a factor for wind pressure.

Equating the heeling and righting moments at 20° for the 30-footer, we have $A \times 1.15 \times 24.6 \times .94 = 24{,}200 \times 1.41$. 24.6 is the distance in feet between the center of effort and lateral resistance; 94 is the cosine of 20°; 24,200 is the displacement in pounds and 1.41 is the righting arm in feet at 20°. Solving the equation, we get for the area 1283 square feet.

Fig. 61 shows wind pressures for various wind velocities.

HOW TO MEASURE AREA

The area of the sail plan is easily found by dividing it up into triangles and finding the area of each by measuring its base and altitude and taking half the product. Fig. 62 shows the manner in which these measurements were taken on the 30-footer. The mainsail was divided

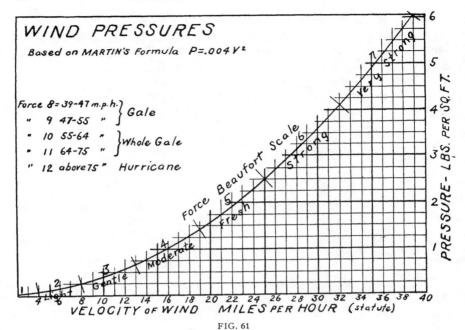

FIG. 61

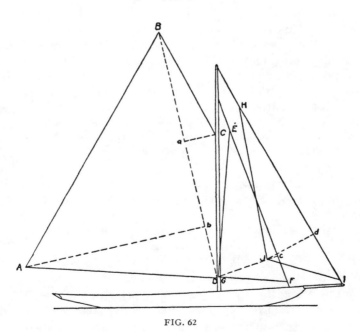

FIG. 62

into the triangles *ABD* and *BCD*, *Ab* and *aC* being the respective altitudes.

The area of the mainsail is then ½ *BD* (*Ab*+*aC*). The altitudes of forestaysail and jib are *Gc* and *Jd* and their areas are

$$\text{Forestaysail} = \tfrac{1}{2}\ EF \times Gc$$
$$\text{Jib} \qquad\quad = \tfrac{1}{2}HI \times Jd$$

By the center of effort of a sail plan is generally meant the center of gravity of the areas of all the sails. The true center of pressure is considerably ahead of the center of gravity of the sails, and its exact position is practically impossible to determine. The position of the center of lateral resistance is also impossible to calculate. As the positions of these two important points are unknown, it is convenient and customary to use instead the center of gravity of the sail plan and the center of lateral plane, the lead being the distance the former is forward of the latter.

HOW TO FIND CENTER OF EFFORT

In finding the center of effort, the center of each sail is found and the center of the system is then found by taking moments. Fig. 63 gives the

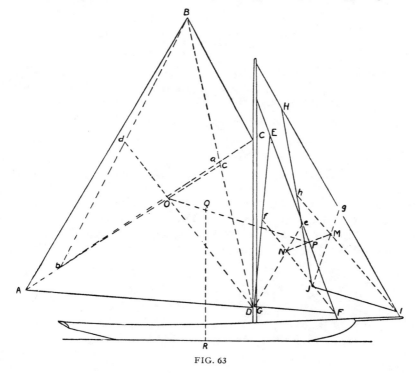

FIG. 63

construction for center of effort of the 30-footer. In finding the center of the mainsail, the second construction given in Chapter II for center of a quadrilateral, with no two sides parallel, was used. (See Fig. 13) *Ab* was laid off equal to ·*aC; bB* was drawn and bisected, its middle point being at *d. BD* was also bisected, its middle point being at *c.* The center of the sail is at *O*, the point of intersection of *bc* and *Dd*. The center of the jib was found by bisecting *HI* and *HJ*, *g* and *h* being the middle points respectively and drawing *Ih* and *Jg;* the

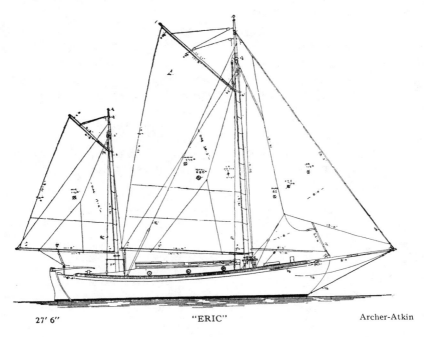

27' 6"   "ERIC"   Archer-Atkin

center is at *M*, their point of intersection. The center of the forestaysail, *N*, was found in a similar manner. The combined center is obtained by finding the center of two sails and then combining with a third, and so on for any number of sails. The areas of the sails for the 30-footer are: mainsail, 934 sq. ft.; forestaysail, 174; and jib, 177. The center of staysail and jib is found at *P* by taking moments about *M* thus:

$$MP = \frac{174 \times MN}{174 + 177} = 3.55 \text{ ft.}$$

Now, taking the moment of the mainsail about *P*, we get the center of all three sails at *Q*, its distance from *P* being

$$PQ = \frac{934 \times PO}{1285} = 16.8 \text{ ft.}$$

The vertical projection of Q on the water line at R gives a convenient point for locating the fore and aft position of the center of effort.

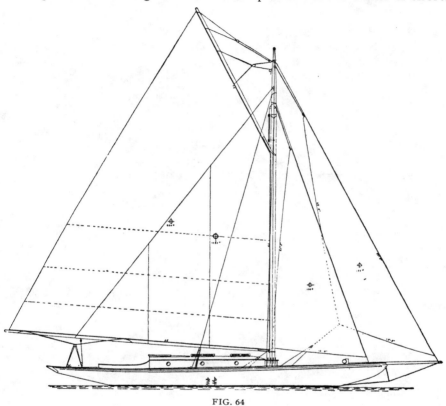

FIG. 64

PROPER LEAD

The proper lead or distance between the centers of effort and lateral plane varies widely with the type of boat. It is conveniently expressed as a fraction of the water line length. For racing machines of the scow type, it varies from .05 to .15, the balance of these boats depending largely on their trim; for shoal, full-ended centerboarders, the lead lies between .07 and .11; for full-ended keel racing boats, the lead is generally a little less. For cruising boats of normal form, the lead is about .06. In the larger sizes, the lead is generally smaller and the center of effort is often placed very little ahead of the $C.L.P.$ the latter

being figured with rudder. The only way of determining the proper
lead is by direct comparison with some well-balanced boat. Even then,
it is difficult to get it right in the case of long, full-ended racing boats,

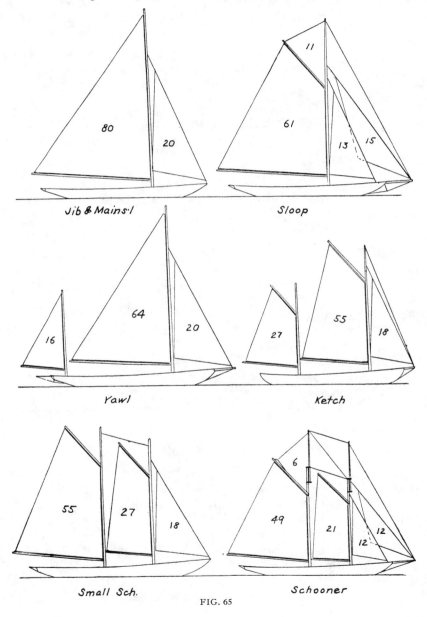

FIG. 65

and the designer generally resorts to the unscientific device of a movable mast. Many designers neglect the topsails when figuring the center of effort, but it is the opinion of the writer that they should always be included in the calculation, as serious defects in balance have resulted from neglecting them.

The area of sail decided upon has to be subdivided into a number of sails in a manner consistent with ease of handling, appearance and efficiency. The principal rigs in use today are the schooner, yawl, ketch and sloop. Fig. 65 shows usual proportions of sail for various rigs and

44' 00"   "TERN III"   Claud Worth

the figures on the sails denote the percentage of the total area. The subject of rig will be discussed in the chapter on sailing cruisers.

JIB-HEADED SAILS

The jib-headed or "leg-o'-mutton" mainsail, to which the term "Marconi" is erroneously applied, has been proved more efficient than the gaff sail by extensive use on racing boats. It is sometimes used on such craft with a curved mast, which is an expensive refinement of doubtful utility. The superior efficiency results not only from the shape of the sail but from the absence of the gaff, which slats about when there is any motion on and throws the wind out of the sail. One might suppose from this that a loose footed sail properly sheeted or one with a boom which could be curved to correspond with the draft of the sail would be superior to the ordinary straight boom. While the rigid boom seems to cause enough bag in the foot of the sail to prevent the wind from spilling downward so that more of its propulsive energy is utilized, the boom which may be bent at will, or which allows the foot of the sail to curve, has certain advantages. The loose-footed sail requires a more rigid boom than the sail with the laced foot and does not seem to work as well in strong winds.

28' 6"   "GOLLIWOG"   Nedwidek

The jib-headed sail is superior to the gaff sail for some types and sizes of cruisers as well. It is easier to hoist, is more easily reefed, and the thrust and chafe of a gaff on the mast are avoided. It does not go too well with the schooner rig, however, as the foresail is apt to be too narrow to stand well. One must forego some advantages in using this rig on large cruisers, for with a gaff one can set a topsail. With slides and a track on the topmast open at the bottom, the topsail may be pulled down in a jiffy, which is the easiest way there is to reduce sail.

GAFF SAILS

With the gaff sail, the sail should be drawn as it is to be cut, bearing in mind that when the sail is set and fully peaked up, the peak angle, or angle between head of sail and axis of mast, is from one to three

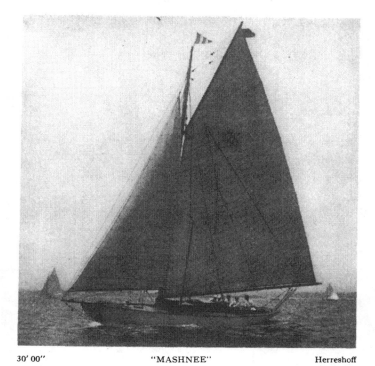

30' 00"   "MASHNEE"   Herreshoff

degrees less than that to which the sail is cut. For a boat without a topsail, the peak angle of the mainsail should not be made less than 27°. This is the peak angle to which the mainsail of the 30-footer is drawn. When topsails are used, the peak angle should be about 42° for the mainsail, and 48° for the foresail in schooners. Low booms and low-cut jibs should be avoided as they are in the way, and are not efficient when the boat is heeled well down. High-pointed jibs are inefficient, as they do not draw well and tend to backwind the mainsail.

CHAPTER X

THE RUDDER

IN DESIGNING the rudder, there are a great many features to be studied and properly coördinated, including: area, shape, amount of balance, location, rake and shape of horizontal sections. The rudder has a most important function but, in performing its useful duty, it occasions

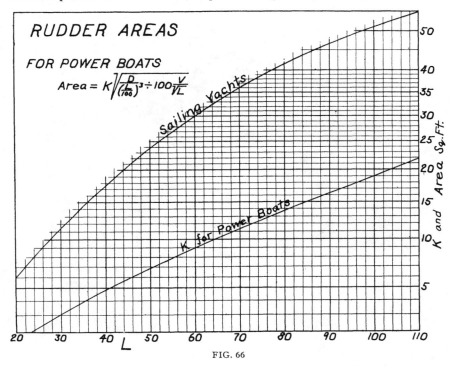

FIG. 66

increased resistance through eddy-making and friction. The problem is to secure adequate steering qualities with the minimum of power losses.

SAILING YACHT RUDDERS

Fig. 66 gives suitable areas of rudder for keel sailing yachts of normal proportions, vessels whose beam, draft, displacement and sail area conform closely to values given by curves in Figs. 23, 24, 25, 53, 54 and 60. Racing yachts may do with somewhat less area than indicated.

100

Sailing yacht rudders should be unbalanced — that is, the axis of rotation should be at the forward edge and they should be hung on the deadwood. Racing yachts formerly were sometimes fitted with balanced rudders located well aft, as shown in Fig. 67, on the theory that the greater leverage would make them as effective as the stern post location, with less area. It was found, though, that interference with the streamlines from the after edge of the fin caused undesirable steering effects and increased resistance. On shoal draft boats, the rudder is necessarily far aft and sometimes twin rudders are fitted on light, beamy racing boats.

The raking rudder post gives the best shape of lateral plane and neutralizes the lifting component which a rudder with vertical post has when yacht is heeled. The best shape gives a fair, symmetrically curved after edge with center of area near mid-depth. A little width at the top, as in Fig. 68, fairs the rudder post as it emerges from the hull, preventing eddies at that point.

The shape of the horizontal sections is important, particularly in racing boats. The rudder should fit snug to the deadwood and offer the least possible break in the surface at that point. It should thin down to almost nothing at the after edge. With a rudder post of fair diameter, the rudder may advantageously be shaped to be a continuation of the streamlines of the hull. Fig. 69 shows such a rudder on a racing boat. Fig. 70 shows horizontal section of rudder on a cruising boat, with metal rudder stock. Fig. 71 shows similar section of fisherman style rudder.

The rudder is a pressure surface similar to the fin or centerboard and the remarks made in regard to lack of efficiency of these members through enforced symmetry of horizontal sections apply to the rudder as well.

POWER BOAT RUDDERS

Average areas for rudders for power boats are found by taking K from Fig. 66 and multiplying it by the square root of the displacement-length ratio divided by 100 times the speed-length ratio. Where the displacement-length ratio is 100 and the speed-length ratio is 1, K is the area, otherwise the correction increases the area for heavier displacements and decreases it for higher speeds.

The rudder should be abaft the screw, for in that position it gets the benefit of being in the column of water thrown aft by the screw and exerts some turning effect even before the boat gathers headway. If placed ahead of the screw, it not only loses a great deal of the influence of the accelerated propeller race but directs an oblique stream on the

upper sector of the propeller which causes the latter to throw the stern
the wrong way.

Power boat rudders may be balanced to advantage, that is, hung
with the axis abaft the leading edge so that not over one-fifth the area
is forward of the axis. If the axis is farther aft than this, it is apt to
coincide with the center of pressure, or even be abaft it, which causes
the rudder to tend to take an angle of itself.

Rudders entirely without external supports are best for resistance, as
skegs and rudder posts cause eddies. This applies to single screw boats.

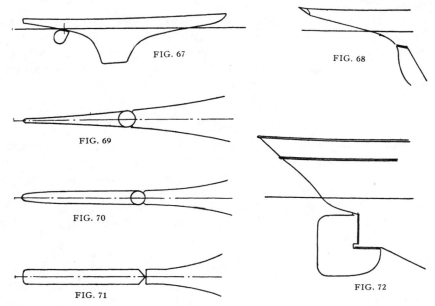

FIG. 67

FIG. 68

FIG. 69

FIG. 70

FIG. 72

FIG. 71

For twin screw boats, the rudder may be supported and balanced to
good advantage, as in Fig. 72. Thick streamline rudders are used with
high efficiency on large ships but as speeds increase, relative thickness
must be reduced.

TURNING CIRCLE

The maximum effective helm angle is about 35°.

When the helm is thrown hard over, the stern is thrown off to one
side as shown in position 2, Fig. 73. The boat then commences to move
in a curved track until she is in position 3, where she is traveling
uniformly in a circular orbit. P, called the point of gyration, is the
point where the center line of the vessel is tangential to the turning
circle. It will be noticed that this point is well forward of the center of

the vessel and that the vessel is really traveling at an angle with the direction of motion, which, of course, causes greatly increased resistance and consequent diminution of speed.

BOW RUDDERS

If steering were accomplished by a bow rudder, P would be abaft amidships, and if by bow and stern rudders — working in opposite

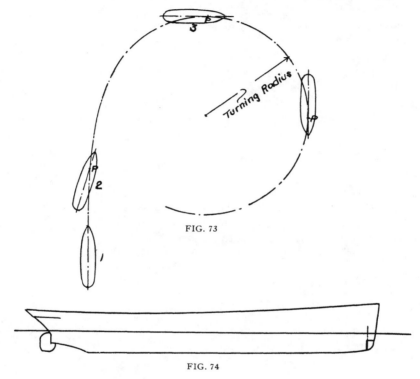

FIG. 73

FIG. 74

directions, of course — exerting equal influences, P would be amidships and the vessel would travel on the circle in the position of least resistance. On this account it would seem that there would be an advantage in fitting a small rudder in the bow, as shown in Fig. 74, to be connected up with the stern rudder to turn equal angles in opposite directions. Another advantage of this arrangement is that the stern is thrown less to one side when making a quick turn, thus decreasing the danger of collision at the stern when dodging a vessel or other obstruction in close quarters.

Mr. Herreshoff designed and built some small double-ended ferry boats with rudders at each end, which, unlike rudders on ordinary ferry boats, were fitted to swing 360°. Ordinarily the forward rudder swung freely but in an emergency could be swung in the opposite direction from the stern rudder, thus giving a smaller turning circle. When entering the ferry slip across a tide, it could be turned the same way as the rudder aft, thus setting the boat bodily to one side.

HYDROPLANE RUDDERS

Bow rudders are used in step hydroplanes, as when the boat is planing a better leverage is obtained in this position. The pressure is exerted on the opposite face from a stern rudder and a deep rudder would exert a capsizing rather than a righting moment. The rudder post should be inclined slightly forward at the top to avoid any depressing tendency.

TURNING MOMENT

The pressure on a rudder in pounds is:

$P = 1.5 \times Area \times V^2$ (ft. per sec.) $\times$ sine of rudder angle.
For 35°, $P = .86\ AV^2$.

The speed should be taken as 10 per cent greater than the speed of the boat with single screw to allow for the propeller race. The center of effort may be assumed for rectangular rudders to be .4 the width from the forward edge, and the distance from the axis of rotation to the center of effort is the turning arm. Then the twisting moment in inch lbs. is $T = P \times Arm$ (in inches).

SIZE OF SOLID RUDDER STOCK

Where rudders are supported at the heel and the stock has torsional stress only, the diameter of the rudder stock in inches is found by the following formula:

$$\text{For solid stocks, } T = \frac{\pi}{16} f d^3 = \frac{f I(polar)}{y}$$

(*Polar* $I = zI$ as set forth in fig. 17)

$$\text{For hollow stocks, } T = \frac{\pi}{16} f\, \frac{D^4 - d^4}{D}$$

Values of f are given in Fig. 17, page 27.

For an example, take the rudder of a 50-footer with Tobin bronze rudder stock, speed 10 knots, area of rudder 7 square feet, arm 13 inches:

$$P = .86 \times 7 \times \left(\frac{1.1 \times 10 \times 6080}{60 \times 60}\right)^2 = 2080 \text{ lbs.}$$

$$T = 2080 \times 13 = 27000 \text{ in. lbs.} = \frac{\pi}{16} \times \frac{60000}{4} \times d^3$$

Solving, $d = 2.09$ in. with safety factor of 4.

DEEP RUDDERS

Where the rudder has no support outside the hull, it is natural to make the blade wide and shallow which brings the center of pressure high up and reduces the bending stress on the stock. For narrow, shallow hulls there is an advantage in making the rudder narrow and deep as shown in Fig. 75. This brings the center of pressure of the rudder lower than the center of lateral resistance of the hull, setting up

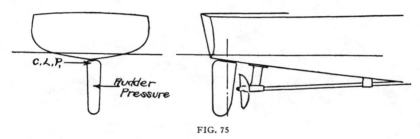

FIG. 75

a couple when the helm is hard over that tends to heel the boat toward the inside of the turning circle. This combats the tendency of the boat to heel the other way from centrifugal force and results in the turn being made on a substantially even keel.

IMPROVEMENTS IN RUDDERS

The ordinary rudder has come down to us practically unchanged through many centuries and has served its purpose well. A number of ingenious rudders embodying distinct improvements have been invented during the present century, among which are the Flettner rudder shown in Fig. 76. The main rudder swings freely through 360° and is turned by the action of the small auxiliary rudder fitted in the after part of the blade. The area of this small rudder is about 1/20 of the area of the main rudder and it is controlled by horizontal rods connecting it with a yoke on a vertical shaft inside the hollow rudder post. When the auxiliary rudder is turned to port it swings the main rudder to starboard, which swings the stern to port. The principal advantage of the Flettner rudder is in the saving of power to operate, as it requires less than 1/10 the power to turn the small rudder as it would the main rudder, making it feasible to steer large vessels entirely by hand. Secondary advantages are elimination of torsional stress in

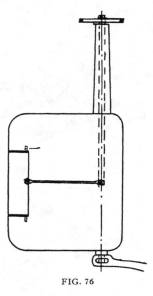

FIG. 76

main rudder post and steadiness in steering. In a seaway, with helm held steady, the main rudder swings back and forth under wave pressure, the ship meanwhile holding a steady course; with the ordinary rudder control it is necessary to swing the helm back and forth to keep the ship steady.

CHAPTER XI

SPARS AND RIGGING

HAVING designed a good hull — the resistance factor in our problem of propulsion of the sailing yacht — we have next to consider the motive end of it, the rig, a problem of equal importance. Type of rig, shape and area of sails are discussed elsewhere and this chapter will deal with stresses in spars and rigging and the problems of designing of these parts to perform their functions properly.

CONSISTENT STRENGTH

It will be admitted readily that it is poor designing to construct a chain with links of various strengths. So, too, in a yacht's rig it is poor designing to make some spar or portion of the rigging much stronger than the rest of the outfit. If a part is stronger than necessary it is pretty sure to be heavier than is required and there is no better place to save weight in a yacht than aloft. Theoretically, shrouds, spars and rigging should be of such uniform strength as to fail simultaneously under sufficient stress, like the "one hoss shay," and the capacity of the yacht engineer is nowhere more clearly indicated than in proportioning the various elements of the rig. Here is where the amateur and the artistic genius are apt to fall down. Nothing will take the place of a thorough knowledge of the strength of materials and of the methods of determining stresses.

DETERMINATION OF STRESSES

A discussion of the theory of beams, columns, elasticity, etc., would be out of place in this book but a brief outline of the application of "graphic statics" to our problem will be useful. Briefly, if we know the amount and direction of a system of external forces and the direction of the resisting elements of the structure (internal forces) we may plot a diagram in which each line is parallel in direction to a given force or resisting element and, being drawn to scale, the lengths of the lines will indicate the amount of the forces. A few simple examples will make the process more or less clear.

In Fig. 77, we have a weight of 5 lbs. attached to a pulley on a slack cord. The stresses on the cord on both sides of the pulley will obviously be the same. Using what is known as Bow's notation we place a letter on each side of each force, as *A*, *B*, *C*. Then *AB* and *BC* represent the

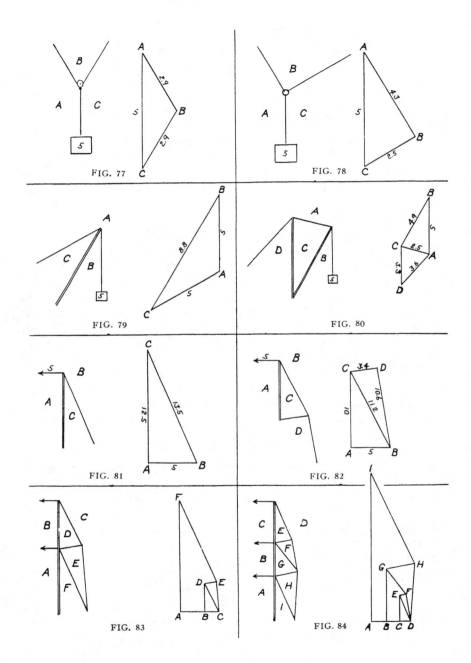

FIG. 77

FIG. 78

FIG. 79

FIG. 80

FIG. 81

FIG. 82

FIG. 83

FIG. 84

unknown stresses in the cord and *AC* the known force of 5 lbs. We then construct the stress diagram at the right, first drawing a line 5 units long parallel to the cord the weight is hung on, lettering one end *A* and the other *C* to correspond to the letters on either side of the cord; then draw *CB* parallel to the right hand portion of the slack cord, and *AB* parallel to the left hand portion. Then the length of *AB* and *BC*, measured in the same units as *AC*, gives the amount of the stress in them. Note that if the paper is held so that a given line in the left hand diagram is horizontal, the letter which is above the line will be at the right hand end of the corresponding line in the stress diagram if the force is tension, and at the left for compression, as for instance *BC* in Fig. 77, which is in tension.

Fig. 78 shows a 5 lb. weight hung from a ring supported by two cords of different angles with the vertical. This brings unequal stress on the two cords. Proceeding as before, we lay off *AC* = 5 and drawing *AB* and *CB* parallel to the supporting cords we find, by scaling them on the stress diagram, the exact stresses in the supporting cords. This is the procedure, always to lay off the known forces in the stress diagram and then complete it by drawing lines parallel to the unknown forces to complete the diagram, and then measuring the length of these lines. If the structure is in equilibrium, the lines will come together to make a closed figure.

Figure 79 shows a weight *AB* hung from a slanting post *BC* that is supported by a guy *CA*. As before, we lay off the vertical line *AB* for the known force and draw *BC* and *CA*, the lengths of which give the thrust in the post and the tension in the guy respectively. This problem is similar to that of determining thrust in a gaff and tension in peak halliards.

Fig. 80 shows a weight hung from a derrick and the construction of the stress diagram will be apparent. *BC* represents the thrust in the arm, *CD* the thrust in the mast and *CA* and *DA* the tension in lift and guy.

Fig. 81 shows method of determining the thrust in a bowsprit and the tension in bowsprit shroud. *AB* being the horizontal component of the wind pressure on the jib. This case is similar to that of a mast with single shrouds.

Fig. 82 shows stresses on a bowsprit and shroud when there are bowsprit spreaders, *CD* being the thrust on the spreader.

DIVISION OF PRESSURE AMONG PANEL POINTS

Fig. 83 shows stress diagram for a mast with upper and lower shrouds and a single pair of spreaders. Here we have two loads, *AB* and *BC*. Of course, the wind pressure on the mainsail is distributed all along the mast but we divide it into three parts, applying the portions at the deck and where each pair of shrouds is attached to the mast. We do not know just what the distribution of pressure is along the mast but we do know that the pressure per square foot is greater on the upper portion of the sail than that on the lower portion, so that we assume the equivalent shape of sail for uniform pressure to be about as indicated by the dotted lines in Fig. 85. Then, dividing the areas between panel points in halves, we consider this pressure on the area $ABCD$ to be applied at the upper shrouds, and that on $BEFC$ at the lower shrouds. The jibstay lands at or near one of the panel points, so half the total pressure on the jib is taken as coming at that point.

With three pairs of shrouds, the stresses work out as in Fig. 84. If any of these shrouds are double the stress is, of course, less in each

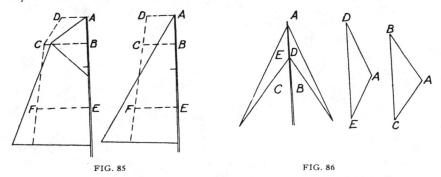

FIG. 85   FIG. 86

though considerably more than half if they are spread. EC gives thrust on the upper portion of mast, GB on the middle portion and IA on the lower portion.

So far we have considered only the stresses on shrouds due to pressure on the sails. There are other important stresses which produce thrust on the mast. These result from the tension on the jibstay from the pressure on the jib, the pull from the main sheet, the weight of the sails, booms and gaff and the pull of the halliards. It has been found that the tension on the head stays (it is about equally divided if there are two) is generally about the same as that on the lower shrouds. Having found this, we can work out the thrust on the mast, as in Fig. 86. The pull in stays AD and AB is balanced by AC and AE, which apply at the deck at the point of attachment of the main sheet. The thrust on upper part of mast from this source is DE, and on lower part $DE+BC$. To get the total thrust on mast, we add the thrust due to the pull of sheet and stays obtained as above to that resulting from lateral pressure obtained as in Figs. 83 or 84, plus the weight of sails and small spars, plus the pull on the halliards. This pull is about 1.5 times the weight of sails and small spars for single part halliards, and less if the halliards have a purchase aloft, according to the number of parts in the purchase.

SIZES OF PARTS

Having found the loads on spars and rigging, the remainder of the problem is to proportion the parts to withstand the stresses on them.

TABLE IX
STRENGTHS OF WIRE, CHAIN & ROD

DIAM.	STANDING RIGGING — BREAKING TENSILE STRENGTH				RUNNING RIGGING — BREAKING TENS. STR.			CHAIN — PROOF TEST		RODS — BREAKING STRENGTH			AREA
	IRON Double Galv.	CAST STEEL	PLOUGH STEEL	AIRCRAFT STRAND 19 Wire	C. STEEL TILLER ROPE	P. STEEL TILLER ROPE	P. STEEL EXTRA PLIABLE	BBB CLOSE LINK CHAIN	BB OPEN LINK CHAIN	NICKLE STEEL Tension	MILD STEEL Tension	MILD STEEL Shear	
1/16"				500 Lb				Proof Test = ½ Breaking Strength					.00307
5/64"				780									.00479
3/32"				1100									.0069
7/64"				1600									.0094
1/8"				2100									.0123
5/32"				3200									.0192
3/16"	1220 Lb			4600					860 Lb				.0276
7/32"	1480			6100									.0376
1/4"	1980			8000	1.35 Ton	1.55 Ton			1530	2.45 Ton	1.47 Ton	.98 Ton	.0491
9/32"	1.2 Ton	2.5 Ton	3.4 Ton	5.0 Ton									.0621
5/16"	1.42	3.5	4.4	6.25	2.0	2.25			1.20 Ton	3.85	2.31	1.54	.0767
3/8"	1.95	4.6	5.9	8.75	2.75	3.15	5.1 Ton	2.24 Ton	1.73	5.50	3.30	2.20	.1104
7/16"	2.35	5.5	7.0	11.75	3.7	4.1	7.2	3.26	2.85	7.50	4.50	3.00	.1503
1/2"	3.39	7.7	10.0	14.25	5.3	5.8	9.3	3.90	3.08	9.80	5.88	3.90	.1963
9/16"	4.46	10.0	12		6.5	7.2	11.5	5.6	3.9	12.4	7.4	4.90	.2485
5/8"	5.7	13.0	16		7.9	8.6	14	6.7	4.8	15.3	9.2	6.14	.3068
11/16"		15.4	18							18.5	11.1	7.40	.3712
3/4"	7.8	18.6	23		11.4	12.8	21	9.5	6.9	22.1	13.2	8.80	.4418
13/16"	9.4							11.2	8.1				.5185
7/8"	11.1	24.0	31		15	16.5	27	12.9	9.4	30.0	18.0	12.0	.6013
1"	14.1	31.0	39		20	22.5	35	15.7	12.3	39.2	23.5	15.7	.7854
1⅛"	18.0	37.0	47				44	20.7	15.0	49.7	29.8	19.8	.9940
1¼"	23.0	46.0	60				55	25.2	18.3	61.0	36.9	24.6	1.227
USE SAFETY FACTOR	5*	5	5	5	6	6	6	3	3	5	5	5	
ELASTIC LIMIT	.8	.8	.8	.8						.5	.5	.5	

STRENGTHS GIVEN IN TONS OF 2000 LBS.

STRETCH OF WIRE = .8 % AT ELASTIC LIMIT

Strength of new 3 strand Manila Rope = ⅓ that of Iron Wire

Starting with the tension members, let us consider first the bobstay. This may be of wire, chain or rod, according to circumstances. It is not important to save weight in this member as it is low down and windage is not a factor. Table IX shows strengths of various grades of wire, chain and rod. From this table it will be noted that there is considerable variation in the strengths of the various grades. If wire is to be used for the bobstay, some designers prefer iron wire since it is double galvanized and will withstand the severe conditions it is subjected to better than the higher strength wires which are thinly galvanized. Its greater weight is not objectionable. Stainless steel wire is growing in favor for such use. Proper factors of safety are indicated in the table. Wire rigging is generally attached to the hull by a plate, pin and a turnbuckle. The pin is in double shear and the plate is subject to crushing under the pin. Since the crushing strength of metals used is about the same as the tensile strength, and the shearing strength about two-thirds, we have, if the crushing and shearing strengths are to be equal:

$$d \times t \times f = 2 \times \frac{\pi d^2}{4} \times \tfrac{2}{3} f$$

whence $t = 1.05\, d$
where $d =$ diam. of pin
$t =$ thickness of plate
$f =$ ultimate strength of material.

That is, for equal shearing and crushing strengths, the thickness of the plate should be as great as the diameter of the pin. It is not, as a rule, practicable to make plates thick enough to satisfy this condition, as clevises in turnbuckles seldom have a spread as great as the diameter of the pin; nor is it necessary, as a factor of safety of 2.5 does well enough for figuring crushing strength. To take an example, suppose we have a bobstay with an estimated stress of 5000 lbs. and the bobstay plate is to be $\tfrac{1}{4}''$ mild steel (tensile strength 60,000 lbs. per sq. in.) then

$$d \times \tfrac{1}{4}'' \times 60,000 = 2.5 \times 5000$$
$$d = .83''$$

A pin of this diameter in double shear, using a safety factor of 5, would be suitable for a load of

$$2 \times \frac{\pi \times .83^2}{4} \times \tfrac{2}{3} \times 60,000 \div 5 = 8700 \text{ lbs.}$$

The designer must design or select each part — plate, pin, shackle, turnbuckle, stay, etc. — in the series to bear the computed load, using a suitable factor of safety to allow for stretch, deterioration, defects and unusual stresses. This is of great importance as there is apt to be a weak unit if every part is not carefully figured.

Remember that weight low down is not detrimental and that the designer can afford to use a higher factor of safety for parts on deck and below than those aloft.

The sizes of the shrouds are calculated in the manner described for the bobstay. Here iron wire will not do, for the extra weight for the necessary strength is prohibitive, to say nothing of the clumsy appear-

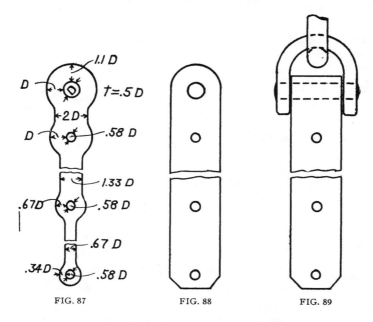

FIG. 87   FIG. 88   FIG. 89

ance of such heavy rigging aloft. Many modern yachts use stainless steel wire rope for all standing rigging. At the deck, the shrouds usually splice into turnbuckles which shackle to chain plates.

Fig. 87 shows a chainplate of uniform thickness designed for uniform strength and minimum weight. Chainplates, unless they are cast, would not be of such a complicated shape. They are generally made from a strip of plate of sufficient width, as in Fig. 88 or Fig. 89. The latter is the type of plate used on the larger yachts and fishing vessels. With it there is no question about the crushing strength over the pin. The pin must be carefully computed for adequate shearing strength. The shackle insures a true lead for the shroud. The head of the plate in Fig. 88 must be bent inboard so as to be in alignment with the shroud, else the pin will not bear squarely and the clevis of the turnbuckle will have a bending stress.

Shrouds aloft are generally spliced around the mast and supported by shoulder cleats. Splicing, if carefully done, does not weaken the wire but corrosion is apt to start in a splice on account of the galvanizing having been chipped off. Where the mast is fitted with a track, as is necessary with the jib-headed rig, eyesplices interfere with the track being fitted snugly to the mast and some other method of attachment of the shrouds is preferable. They may be shackled to an eyebolt passing through the center of a hardwood plug through the mast or into an eye screwed into a piece of stout tubing passing through the mast. The bolt or tube should be supported on each end by a plate well screw-fastened to mast. Plate fittings, or "tangs," screwed to the spar are much used and are superior to the above.

Bobstay Plates

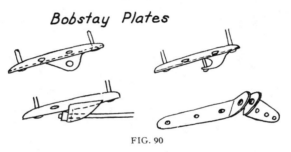

FIG. 90

Wire halliards offer some decided advantages for cruising as well as racing yachts. They do not stretch and allow the sail to sag, they offer small windage and wear much longer, an important advantage for yachts that do extensive cruising at sea. Muhlhauser, who completed a round-the-world voyage in the 52-foot w.l. cutter *Amaryllis* in the summer of 1923, stated that his wire halliards lasted 2 years but that his manila halliards required replacement every 3 months. The Austrian ketch *Sowitasgoht*, which crossed the Atlantic the same summer, was equipped with a wire halliard leading to a two-speed winch on the mast near the deck. As the sail was hoisted, all the wire was reeled up on the winch, a neat arrangement for stowing the slack and one which does away with the usual complication of having the hauling end of manila spliced into the wire and sheaves aloft wide enough to take the manila.

COMPRESSION MEMBERS

All the spars are in more or less compression. The booms have the least of all, except the spinnaker boom, which is in pure compression as are the spreaders, if of the swinging variety. The mast, if thoroughly stayed, is in almost pure compression and in figuring its strength we may ignore the comparatively slight bending stresses. The breaking strength of long columns is, by Euler's formula (see Fig. 17),

$$W = \left(\frac{\pi}{L}\right)^2 EI$$

W is in pounds, and L in inches.

Let us work out the strength of a round solid mast 3″ diameter and 17½′ long as an example of the application of this formula,

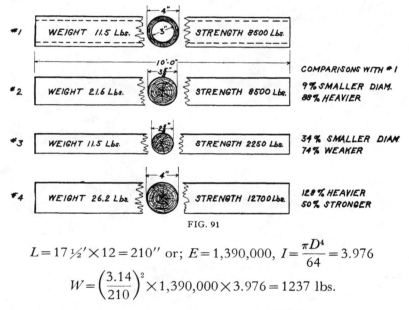

FIG. 91

$$L = 17\tfrac{1}{2}' \times 12 = 210'' \text{ or; } E = 1{,}390{,}000, \; I = \frac{\pi D^4}{64} = 3.976$$

$$W = \left(\frac{3.14}{210}\right)^2 \times 1{,}390{,}000 \times 3.976 = 1237 \text{ lbs.}$$

This is for two pin ends but this case is for one fixed end so the strength is 2474 lbs.

With a factor of safety of 3.5, the mast will bear a load of $2424 \div 3.5 = 707$ lbs.

Any shaped section may be calculated in the same way. For instance, take a hollow rectangular section 3″×4″ outside and 2″×3″ inside, 17.5′ long. (See the fourth section in Fig. 17.)

$$I = 11.5$$

$$W = \left(\frac{\pi}{210}\right)^2 \times 1{,}390{,}000 \times 1.15 \times 2 = 7160 \text{ lbs.}$$

SPRUCE
I	2.718
E	1.39 MIL.
EI	3.78 MIL.

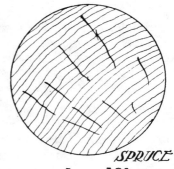

SPRUCE
I	3.98
E	1.39 Mil.
EI	5.53 "

DURAL
I	.601
E	.10 Mil.
EI	6.01 "

DURAL
I	.121
E	10 Mil
EI	1.21 "

STEEL
I	.208
E	30 Mil.
EI	6.24 "

STEEL
I	.0161
E	30 Mil.
EI	.48 "

SPRUCE	DURAL
I 1.91	.208
E 1.39 Mil.	10 Mil.
EI 2.66 "	2.08 "
	2.66
	2.08
EI 4.74 Total.	

THE SECTIONS IN EACH COLUMN ARE SAME WEIGHT PER FOOT

FIG. 92

This shows what an enormous increase in strength we get with a little increase of moment of inertia. Conversely, by hollowing a mast, which reduces the radius of gyration slightly, a great deal of sectional area and, consequently, weight may be saved. This point is well illustrated in Fig. 91, where the hollow spar at the top is compared as to diameter,

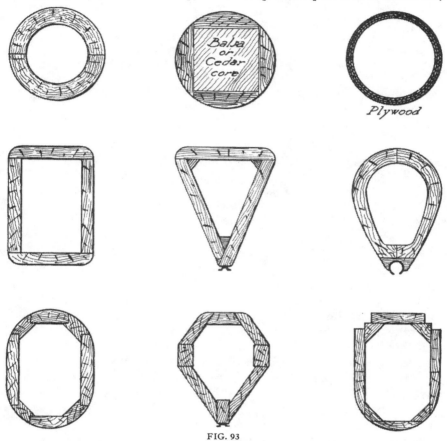

FIG. 93

strength and weight with various solid spars of equal length, having respectively the same strength, weight and diameter as the hollow spar.

Figure 92 compares the strengths of spruce, duralumin and steel spars of equal weight in an interesting manner. The hollow sections are 3″ outside diameter. The value EI is a measure of the strength from which we see that the steel spar is the strongest. Where there is no limit in outside diameter, the lighter material shows up best as indicated by the solid sections at the right.

Hollow spars are of obvious advantage in racing boats but they are almost as desirable for cruising boats. Light spars are more easily handled and greatly reduce the wear and tear on hull and rigging, especially on long voyages.

Hollow spars have been made in a great variety of sections, some of which are shown in Fig. 93. If the spar is supported in all directions at

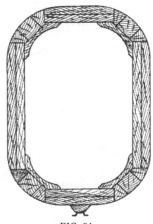

FIG. 94

both ends, the greatest strength is obtained by using a section having the same radius of gyration about both axes; if it is supported at one end in only one direction, then an oval or oblong section is preferable. Such a case is the common one of a mast with three pairs of shrouds attached to it at three points. The intermediate and upper shrouds are attached at points where the mast is stayed fore and aft by stays so that at these points the mast is held in all directions but the lower shrouds are the only support to the mast at their point of attachment. The mast, therefore, is free to bend in the fore and aft plane at this point and a mast section that is deeper fore and aft than transversely helps to correct this condition.

A mast supported as described above may be considered to be a column having a fixed end at the deck and at the intermediate shrouds and a pin end at the upper shrouds. The strength of such a mast should be figured for the section from the deck to the intermediate shrouds, taking the radius of gyration of a section at midheight about a transverse axis, as the mast is free to bend in the fore and aft plane only since it is supported transversely by the lower shrouds. The load is the maximum load on the mast, and the correction factor is 4. The upper part of the mast is much more lightly loaded and should be figured separately, using the least radius of gyration of a section midway between the shrouds. The correction factor is 2. The section of a spruce mast for a Class M yacht is shown in Fig. 94.

The process in figuring strength of struts is one of trial and error, by drawing a section that looks right and calculating the strength, then correcting the section and refiguring until a section is obtained that gives just the necessary strength.

Normal proportions of the usual cruising yacht spars are shown in Fig. 95. In this diagram, lengths of masts are measured above the deck and maximum diameters are given as fractions of the length, and other diameters as fractions of the maximum diameter. The diameters found in this way are nearly correct and will serve as a guide and a starting point for calculating strengths. The loading of booms is almost impossible to determine and reliance may be placed on the proportions indicated in the diagram. If the boom or gaff is to be made hollow, a section may be used having the same ratio as that of a circle of the diameter indicated for the solid spar.

$$\frac{L}{\sqrt[3]{\dfrac{I}{Y}}}$$

STEEL SPARS

The maximum compressive load to be borne by steel masts may be calculated by Euler's general formula just as we did for spruce,

$$P = \frac{\pi^2 EI}{L^2}$$

For plain hollow steel cylinders this works out

$$\text{Breaking load} = \frac{14{,}800{,}000\ (D^4 - d\ )}{L^2}$$

Where D is outside diameter in inches
Where d is inside diameter in inches
Where L is length in inches.

UNSUPPORTED SPARS

Where masts are entirely unsupported by stays or shrouds, we have bending stress only and we may treat it as a cantilever uniformly loaded. The diameter in inches may be calculated by the formula

$$d = \sqrt[3]{\frac{16\ PL}{\pi f}}$$

where *P*, the load, is the total pressure on sail, *L* the length of the mast in inches and *f* the maximum fibre stress, values of which for various materials are given in Fig. 17.

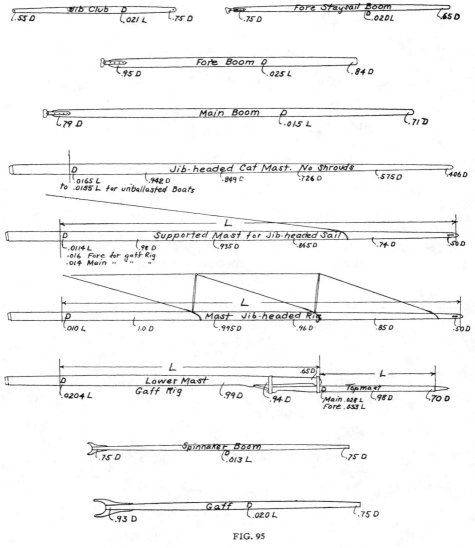

FIG. 95

Suppose we calculate diameter of a spruce mast 20 feet above the deck with 200 square feet of sail:

$$d = \sqrt[3]{\frac{16 \times 4 \times 200 \times 20 \times 12}{3.14 \times 5000}} = 5.80''$$

The 4 was introduced as a factor of safety and the wind pressure was taken as 1.

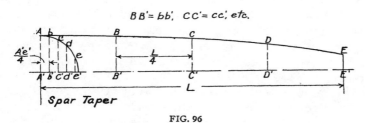

FIG. 96

SPAR TAPER

In designing spars, it is best to draw them much foreshortened to exaggerate the taper. For small yachts, spars may be drawn full size in diameter and with lengths on the scale ½" or 1"=1'. Having determined the maximum and minimum diameters of a circular spar, the proper taper may be laid out as indicated in Fig. 96 which requires no explanation.

CHAPTER XII

CRUISERS

DESIGNING the ideal cruiser is a most complex problem, for the attainment of many virtues, some of them antagonistic to one another, is sought. Among this collection of virtues are comfortable living accommodations, safety, speed, ease of handling, durability and economical maintenance. Here, as always in naval architecture, the successful vessel is a happy compromise of many conflicting characteristics.

The evolution of the cruising yacht has been strongly influenced by various commercial and naval types as well as by the racing yacht — all of them types that occasionally have to combat conditions where their ability is tested to the utmost. The performance of yachts that make especially noteworthy deep sea voyages is another important influence in the arrangement, rig, model and construction of the ordinary cruiser.

Among recent long voyages of note are those of the yawl *Amaryllis*, owned by G. H. P. Muhlhauser, which sailed from England in September, 1920, and arrived home in July, 1923, having gone around the world, a distance of 31,100 miles. This vessel is of the English cutter type and was 41 years old when she finished her voyage, having been built of oak, teak and elm to last a lifetime. She is 62 feet over all, 52 feet water line, 13 feet beam and 10 feet draft. An account of her voyage has appeared in book form. Other notable voyages in 1923 include that of the 31-year-old English cutter *Firecrest* and the ketch *Sowitasgoht*. *Firecrest* is only 30 feet on the water line and was sailed singlehanded by her owner, Alain Gerbault, on her voyage from Cannes to New York.

The ketch *Sowitasgoht* is 42 feet over all and was sailed across the Atlantic from Germany by her three owners, who also designed and built the boat. She is of a much more modern type than the two previously mentioned, having moderate overhangs and a distinct appendage. Her rig, also, is distinctly modern, being the jib-headed type and each sail having a wire halliard.

The ketch *Typhoon* made a notable voyage in 1920, sailed by her owner, William W. Nutting, who wrote a most interesting and instructive account of the voyage in his book entitled *The Track of the Typhoon*.

In 1921 the fisherman type schooner *Lloyd W. Berry* made a similar voyage which one of the owners, Mr. Roger Griswold, wrote up in a very interesting manner for *Yachting*. This vessel is 60 feet over all and 47 feet water line. In the account Mr. Griswold points out many interesting lessons learned from the voyage. It is such voyages as these that try vessels more thoroughly than a lifetime of coasting and one may get much useful data for the ideal cruiser from studying accounts of such voyages. The cutter *Neith* and the schooner *Diablesse* also crossed the Atlantic about this time. Still more recent world girdling voyages are those of the *Islander* and the *Svaap*.

An interesting book which every enthusiastic cruising yachtsman should read is *Yacht Cruising* by Claud Worth. Mr. Worth has a wealth of experience in cruising about the British Isles and Europe and this book is full of extremely valuable information and advice on all phases of cruising.

COMFORT

One of the most essential characteristics of the ideal cruiser is comfort, as the cruiser is a nautical dwelling, not merely a means of transportation. Unless one may live aboard with approximately the comforts of home, the zest of cruising cannot endure. This means that he must be tolerably well fed, have a comfortable place to sleep and living quarters that are warm and dry.

No set scheme of internal arrangement can be laid down as ideal, as arrangement is essentially an expression of the owner's individuality and will vary with personal requirements.

The arrangement of Mr. Worth's cutter *Tern III*, shown in Fig. 97, is interesting, representing as it does the ideal of an experienced yachtsman. His objective was to secure quiet sleeping quarters, isolation from the forecastle by two intervening bulkheads and from the cockpit by one compartment. The large engine room utilizes the after portion of the boat in which the headroom is less than the amount necessary for comfortable living quarters.

Headroom of at least six feet under the beams in the center of the boat is one of the essentials for comfort. In yachts of less than 40 feet water line, it is extremely difficult to get full headroom under a flush deck without a cabin house. The cabin house is structurally undesirable but is a necessary evil on small yachts, and even on yachts up to about 75 feet water line if the extreme after portion of the boat is to be used for living quarters. Of course, the floor must be placed as low as possible in the boat, and in small craft the designer's ingenuity is taxed to provide the necessary minimum floor width and to conceal the inner skin of the boat with useful arrangements such as transoms, buffets, lockers, etc.

One of the essentials of a comfortable cruiser is a commodious galley. This most important department must be equipped with the essentials of housekeeping in a small space — range, ice box, sink, dish lockers and cupboards must all be provided in a convenient and adequate arrangement.

The usual location of the galley in a boat of any size is just abaft the forecastle, the full width of the yacht. On a small craft it is frequently

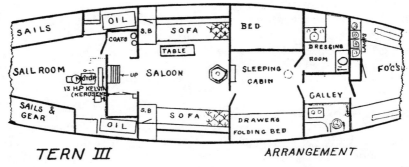

FIG. 97

located in the after end of the main cabin and handy to the cockpit with the best chance of ventilation in rough weather. There is much to be said for this location for very small yachts. A coal range provides grateful warmth and is ordinarily adequate for heating small and medium sized yachts. On larger yachts, it is necessary to provide a special heating arrangement for the owner's apartment, as in the course of extensive cruising much cold and damp weather will be encountered, even in the hottest months. Small stoves of the ornamental Franklin or English tile variety are a simple means of heating and are most cheery.

Ample storage space is another requisite and one that is too often overlooked in the owner's eagerness to get the maximum of living accommodations. Spare gear, supplies and personal equipment require an amount of space hard to realize, even for ordinary coasting, and there is never any too much storage space.

The factors for comfort mentioned above have all to do with the purely domestic side of the problem. There are other vital factors having a decided bearing on the question when the yacht is at sea, such as stability, ease of motion in a seaway, ease of steering, freedom from pounding, etc. Some of these questions have been discussed previously and the remainder will be treated in the following pages.

A staunch cruiser is modeled to negotiate rough water without undue stress, to carry sail in all reasonable winds with slight helm, to steer easily when running before a heavy sea without rooting or yawing and finally, when it is impossible to sail, to be able to "heave to" and ride safely with small sail area, or to the sea anchor. Moderate overhangs are a decided requisite and the forward overhang in particular must be fairly short and with sufficient deadrise to eliminate pounding under all conditions. The after overhang as well should have some deadrise, for under some conditions, as, for instance, when the wind shifts and the yacht is left to contend with a heavy sea travelling in the direction she is sailing, this gets severe punishment. Muhlhauser had an experience of this kind in the Red Sea when it seemed as though the stern would be knocked off. The canoe stern seems to be the best for heavy weather conditions.

Sturdy construction is also of vital importance on the score of safety, although this does not necessitate excessively heavy construction. Good materials properly put together are more effective than a lot of lumber loosely fastened. The fisherman type construction has gained its popularity more on account of its cheapness than for its strength. When such vessels go ashore or get into other difficulties, they generally sustain serious damage, where the scientifically constructed yacht with outside ballast often gets off scot free. Outside ballast, by the way, contributes in important degree to safety, as touched upon in the chapter on ballast.

Ample draft, about the amounts indicated in Fig. 54, is desirable for coasting and seagoing work. If, however, the yacht is to be used mostly out of a harbor where the draft is limited to a certain amount, that in itself is sufficient justification for the shoaler draft and a reduction in safety for seagoing work must be put up with.

The freeboard should be ample, but not excessive. The amounts indicated in the curve, Fig. 55, should not be greatly exceeded. High freeboard is, of course, a great aid in getting the desired headroom and the temptation is strong to raise the freeboard on this account and to justify it further by the theory that seaworthiness is also being increased. This is a mistake, as excessive freeboard raises the center of gravity of the deck, which is an important item of weight, and greatly increases the resistance when on the wind. Some high sided cruising yachts are really poor performers on the wind in a choppy sea, being

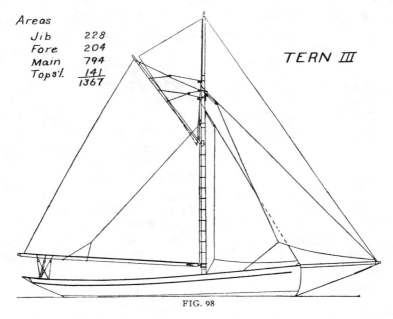

Areas	
Jib	228
Fore	204
Main	794
Tops'l.	141
	1367

TERN III

FIG. 98

seriously handicapped under these conditions by their excessive freeboard.

Durability of construction is extremely desirable, not only for the reason that the vessel is stronger and, consequently, safer but because her resale value is always a much larger fraction of her first cost than is the case with a cheaply constructed boat. English yachtsmen, who make more of a business of cruising than we do, appreciate this and build yachts of oak, teak, elm and mahogany, fastened with bronze and copper, to last a lifetime. Lead and bronze increase the original investment but hold their value and enhance the yacht's value when the owner wishes to sell.

Mr. Worth's yacht *Tern III*, lines and sail plan of which are shown in Figs. 52 and 98, is an excellent example of the English cruising yacht. In model and rig she does not conform to American ideas, but is certainly a very sturdy and wholesome cruiser, and will bear studying. The design was the outcome of much experience and long study to meet rigid requirements, which include cruising about Europe and the British Isles at all times of year. She is much underrigged for American coasting conditions.

RIG

The principal rigs used for cruising are the sloop or cutter, the schooner, ketch and yawl. Each has its virtues and special field of usefulness, so that it is unreasonable to assert dogmatically that any one rig is superior to the others. Mr. Worth's ideas on this matter are worth quoting. He says in his book:

"There is much difference of opinion as to the best rig for an ocean cruiser. Having owned a ketch, two yawls, a schooner and several cutters, as far as my limited experience goes, I am convinced that no rig can equal the cutter, in anything from 15 to 50 tons at any rate. Besides the efficiency of the rig, the lightness of spars in proportion to sail

38' 00"   "JEAN"   Munro

area and the favorable position of the mast near the middle of the vessel, very strong points in its favor, are the simplicity of the gear, the comparative ease with which chafe may be guarded against, and the control one has over the gaff. I have found the ketch the most trying of all in an ocean swell — it is so terribly hard on the gear. To get any useful mainsail area the hoist must be great and the gaff nearly as long as the boom. Owing to the narrowness of the mainsail and the momentum due to the height and weight of the gaff, the spar may slam violently from side to side even when the boom is fixed with rolling tackles. A mizzen, though handy for coast pottering, and useful in a very small craft because it would be instantly available when riding to a sea anchor, is merely a nuisance in a vessel of 20 to 30 tons in open water. There is another point which is generally overlooked. In an ocean cruise the crew must depend entirely upon their own efforts. The only accident to gear which could permanently and utterly cripple the vessel would be the loss of the mast. A cutter's mast can be more efficiently stayed and bears less strain than a mast further forward, and it might be made stout beyond all reason and yet the aggregate weight of spars be less than in a ketch of equal sail area."

THE SLOOP

The virtues of the sloop or cutter have been touched upon above and, in addition, the superior efficiency of the rig is an important consideration. Many yachtsmen scoff at the value of efficiency in a cruiser where one is not supposed to be in a hurry, but an experienced yachtsman appreciates the comfort of making the best possible distance in a long, hard day's thrash to windward and the cutter will greatly outdistance any other rig under such conditions, other things being equal. On account of the cutter's quicker turning tendencies, she should have a more extensive lateral plane than is desirable with the ketch or schooner, where the centers of the individual sails are much farther apart. There is little difference between the sloop and cutter. The cutter's mast is farther aft, being located at about 40 per cent of *L.W.L.* from the bow.

THE YAWL

The yawl rig comes next in speed to that of the cutter, and the ketch and schooner after that. American racing rules rate yawls at 93 per cent of their figured rating, which is a fair index as to their speed as compared with that of sloops. The mizzen of the yawl is stepped abaft the rudder post and is useful as a riding or steadying sail. Its efficiency for propulsion, however, is slight. The back draft from the mainsail interferes with the mizzen drawing when on the wind, even with the sail cut as flat as possible.

The chief advantage of the rig lies in the reduction in length of the main boom, although this is very slight compared with a snug cutter rig with the mast at $\frac{2}{5}$ of the water line from the forward end. The location of the mast forward, however, helps out the cabin arrangement.

THE KETCH

The subdivision of sail is greater than in the yawl. Where the mizzen is large compared with the mainsail there is not so much difficulty in getting it to draw as there is with the yawl. The short main boom is a disadvantage, with the gaff rig, as the gaff tends to slat about in a seaway; with a jib-headed mainsail, it is hard to get sufficient area

31' 6"   "STORMY PETREL"   L. F. Herreshoff

without getting a very high and narrow sail. This rig has the preference of many experienced yachtsmen for deep sea cruising, and in the larger sizes it seems decidedly superior to the schooner.

THE SCHOONER

The schooner rig is certainly popular in America, though not at all so in England. It is difficult to use anything but a gaff sail for the fore-

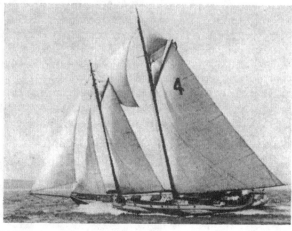

35' 00"   "MALABAR IV"   Alden

sail and have sufficient area, but the combination of gaff foresail and jib-headed mainsail seems inharmonious. The schooner foresail is a poorly shaped sail. With the gaff approximately as long as the boom, it is prone to slat about in a seaway and the sail will not stand as well as one with the gaff short in comparison with the boom. Having the largest sail aft is a poor arrangement for going to sea and running in hard winds. The main staysail rig eliminates the foresail difficulty mentioned above. The reader will infer rightly that the schooner rig is not the favorite of the author.

THE CAT

The cat rig, while it is hardly a seagoing rig, is useful for cruising in sections where the water is shoal. The rig is simplicity itself and on that account a boat as large as five tons is easily handled by one man. The famous Cape Cod cat boats are now nearly extinct, which is to be regretted, as power boats have replaced them for commercial purposes and yachtsmen have turned largely to other types. The Cape cats were really very seaworthy, considering their shoal draft, and were formerly used in great numbers for fishing in the turbulent waters off the elbow of the Cape.

THE MARCONI RIG

The Marconi rig, as it is called in America, but which is more accurately termed the Bermudian, leg-of-mutton or jib-headed sail, has been demonstrated much superior to the gaff rig for racing purposes. It is also becoming popular for cruising boats and has many advantages for this work. It is, in the first place, simple and consequently a cheap rig to install. It is more easily handled and does not slat the wind out as badly as a gaff sail in a choppy sea and light wind. It necessitates taller masts, however, and the use of a track instead of mast hoops. The only real drawback to the taller masts is the greater windage and consequent tendency to drag anchor in a heavy wind; also the fact that in reefing there is no material reduction of top weight. The gaff rig with topsail has the advantage when it comes to reducing sail, as taking in the topsail is equivalent to a considerable reef in a jib-headed sail, and is more quickly and easily done.

EQUIPMENT

This factor is a most important consideration in a cruising boat. Water, stores, boats, ground tackle, heating and lighting arrangements, navigation instruments, etc., must all be provided and arrangements made for the best possible stowage. The amount of water to carry will, of course, depend on the duration of any prospective voyage and the number of people to be carried. Where only short runs are made and it is convenient to take on water frequently, the designer should figure on providing 5 to 10 gallons per person per day. On long voyages with rigid economy, as little as 1 gallon per person per day can be gotten along with. Tin-lined copper or monel tanks are the best containers, with galvanized or painted steel coming next. Hand holes of generous size should be provided to permit inspection and cleaning. Tanks should be well subdivided with swash plates.

Electric light is really the only satisfactory means of lighting and is becoming widely used. Small and reliable generating sets with adequate storage batteries are now standard equipment on the best class of yachts.

The question of boats and their stowage is an important one. Small yachts tow their tenders on coastwise trips but larger yachts, and yachts on deep water, must find a place aboard for them. Davits are a nuisance, except on large yachts, and a good substitute is tackles, one

13′ 6″   "FOX"   Skene

from the cross trees and the other attached to the runners. A pair of wooden chafing guards may be hung over the sides to protect them when hoisting. In a flush deck yacht there is room for the boat on deck, and if a single boat is carried an excellent place for it is to capsize it over a skylight and lash it to ring bolts in the deck. For the small, single-handed cruiser, where a boat to get ashore in the harbor is all that is required, a canoe like the *Fox* may be considered. This little V-bottom craft is 22 inches wide, 13½ feet long and weighs only 26 pounds. It is easily lifted aboard and may be passed below through any companionway. It is extremely easy to paddle, is quite seaworthy and a delightful little boat to get about in.

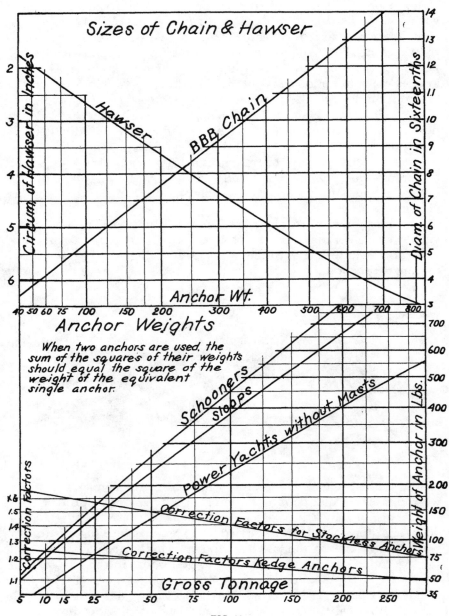

FIG. 99

GROUND TACKLE

Anchors and cable, of sufficient strength and weight to insure the safety of the yacht when it is necessary to anchor in exposed positions, must be provided. Suitable weights and sizes of anchors, chains, and warps are indicated in Fig. 99, which is based on data compiled by N. G. Herreshoff. These curves will be found useful. The weights given are for the Herreshoff anchor, which is particularly efficient. Curves of factors are given by which the weights obtained from the curves are to be multiplied to give suitable weights for the common kedge anchor and for stockless anchors. The weights from the curve are for a single anchor, and if two anchors may be used at once, as is usually the case, the sum of the squares of their weights should equal the square of the weight of the single anchor; thus a 30 lb. and a 40 lb. anchor are equivalent to one 50 lb. anchor.

AUXILIARY POWER

Practically every cruiser nowadays is equipped with power, and its use greatly increases the radius of action and the safety of the cruising yacht. The best location for the engine is an important matter. The best possible arrangement for comfort and safety is to isolate the engine in a compartment of its own, but this is seldom practicable where it is desired to use all possible space for living accommodations.

The engine is sometimes located forward, partly or wholly in the galley, with a portable housing to conceal it and get it out of the way as much as possible. The danger of close association of the galley range with the gasoline engine is not great. Wherever the engine is located it is important to have it readily accessible on all sides and well ventilated.

The engine, if it is to be really useful when the safety of the yacht is at stake, should have sufficient power to drive the yacht at a speed corresponding to a speed-length ratio of at least .8. This means a speed of 4.8 knots in a 36-foot water line boat, or 8 knots in a hundred footer. Many auxiliaries have greater speed than this.

Propeller location is another consideration. A convenient and common location is on the center line in an aperture in the deadwood. It is considered better, however, to run the propeller shaft out on one side of the stern post and then the long, easy water lines in the run are preserved and the efficiency of the lateral plane is not impaired. With a feathering propeller in this location there is less resistance than there is with an aperture cut in the deadwood. With the propeller turning inboard at the top, there is no tendency to carry a helm.

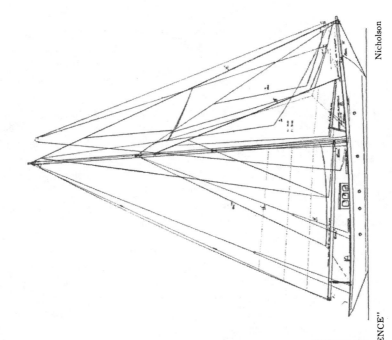

Nicholson

"PATIENCE"

50' 0"

MAINTENANCE

A great deal of expense, labor and upkeep may be saved by minimizing the amount of bright work, and the modern tendency is to eliminate great expanses of bright or scrubbed surface. Decks and spars painted a suitable tint look almost as well as when varnished and are vastly more durable and economical. Mahogany or teak trim about the skylights, rails and companionway, gives sufficient relief to make the craft look yachty. Enameled galvanized steel fittings are satisfactory substitutes for polished brass.

BEAUTY

Beauty is a valuable quality in a yacht, a never ending source of satisfaction to the owner and a tangible asset when he desires to sell. The external appearance must be in accord with the particular class of service for which the yacht was designed, and no matter how severe the conditions which she must meet, it is never necessary to design an ugly boat. Conventionality must be regarded since freak features in a boat greatly lower her resale value.

SPEED

Last, but not least, comes the matter of speed, and it should not be neglected. Speed is always a source of satisfaction and it is sometimes an important factor for safety. There is nothing more disheartening than to try to get to windward smartly in a slow boat when the harbor lies in that direction and weather conditions are anything but pleasant.

POWER CRUISERS

When the craft is entirely propelled by power, the speed at which she is to run completely determines her character. Not only is the external form governed by the question of speed, but the cabin and deck arrangements are also controlled by this factor. When the speed is medium, that is, when the speed-length ratio is 1.0 or less, the machinery weight is inconsiderable and there is considerable latitude possible in the style of the boat and in the location of the engine, which governs to a large extent the internal arrangement. In such a boat it is possible to locate the engine in either end of the craft, leaving most of the space available for cabin accommodations. It is even possible, where a high speed engine is used in a slow boat, to locate it way aft on one side where it connects with the propeller shaft below it by a silent chain drive which reduces propeller speed. The entire cabin space is then available for living quarters.

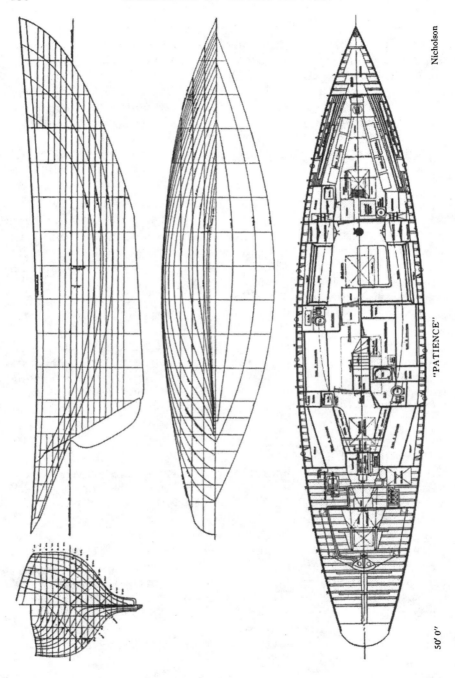

Nicholson

"PATIENCE"

50' 0"

148' 00"

"VEDETTE"

Cox & Stevens

Full powered cruisers have speed-length ratios greater than 1.0. Here the machinery weights are greater and it is desirable to locate the propelling outfit somewhere near amidships, otherwise it is necessary to place weights forward or aft to obtain the proper trim. It is highly undesirable to have the center of buoyancy very far abaft amidships and it is out of the question to have it forward of amidships without serious detriment.

For speed-length ratios of less than 1.2, the fine stern is desirable — that is, one with considerable deadrise — as there is no decided tendency for the vessels to squat at speed-length ratios of less than 1.5.

In moderate-speed cruisers a comparatively heavy and slow speed

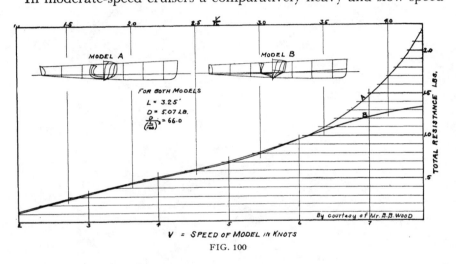

V = SPEED OF MODEL IN KNOTS

FIG. 100

engine is best for reliable performance, since there is little to be gained by extremely light weight up to this speed. Weight can be saved, however, which may be useful for other purposes by using light weight, fast turning engines with gear reduction.

When we get up to speed-length ratios greater than 1.2, we get into the fast cruiser class and light weight machinery is essential. A flat stern commences to be necessary to keep the boat from settling by the stern and something in the way of seaworthiness must be sacrificed.

At a speed-length ratio of over 2.5, the so-called V-bottom type commences to be superior to the round bottom and the phenomenon of planing begins. Figs. 100 and 152 give the results of model tests where the V-bottom and round bottom models of equal displacement were towed and resistance plotted. These curves are instructive and show that the V-bottom type is slightly inferior to the round bottom for medium and slow speeds. The only excuse for its use is its somewhat greater efficiency in the higher speed range. It is little, if any, more economical to build and is inferior to the round bottom type in looks and strength. Through skillful designing the visible characteristics of the V-bottom construction may be made inconspicuous.

Extremely fast cruisers have a speed-length ratio of over 2.5, as indicated in Fig. 20. It is necessary in their case to economize weight in every part of the boat's structure, power plant and equipment. It is also necessary to make great sacrifices in the matter of cabin accommodations and seaworthiness. Great care must be taken in the elimination of all unnecessary wetted surface, in streamlining appendages and in the reduction of wind resistance.

Great improvements have been made in the appearance of power yachts in recent years, the influence of modern naval and commercial vessels being clearly felt. High houses full of large plate glass windows

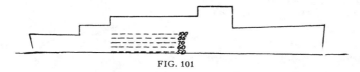

FIG. 101

have given way to compact deck erections with ports or small windows and snug steering positions.

In all but very small sizes of power cruisers it is possible to get full headroom in most of the spaces below without marring the appearance of the boat or making her look topheavy. Designers frequently fail, however, to make the most of their opportunities for securing large accommodations by the use of high deck houses. The modern passenger steamer has an erection above the main deck of considerable height for about two thirds of the vessel's length, and the same general mass can be used in a yacht as shown in outline in Fig. 101.

In Fig. 101 the height of floor is indicated for vessels of 50, 60, 70, 85 and 100 feet length to secure full headroom below the upper deck. In boats of 50 to 60 feet in length, the high house allows the main cabin floor to be raised sufficiently to give useful space beneath it for tanks or storage of gear. From 60 to 70 feet length, this space is considerable and affords room for ice, batteries, lighting sets, luggage and equipment of all sorts. Over 70 feet length this space is useful for low headroom staterooms and engine room, and over 85 feet, for more general living purposes. In boats of 50 feet in length it is possible to get 4½ or 5 feet headroom in the engine room with a saloon or cabin above having full headroom. This amount of headroom in the engine room is good practice, provided adequate ventilation is furnished and a small space having full headroom is arranged near the engine.

The best designing calls for the engines, fuel tanks, lighting set, pumps, batteries, etc., to be concentrated in one compartment near the center of buoyancy, with water tanks forward of and abaft the engine space. It is an excellent scheme to have a water tank way forward and one way aft so that perfect trim may be maintained at all times by drawing from the proper tank. A pair of fuel or water tanks amidships, one on each side, gives a chance also to maintain trim athwartships under all conditions.

It is of great importance to make careful study of ventilation and to provide ample means of ventilating thoroughly all compartments. It is not difficult to do this, but many boats are defective in this respect. With large boats, particularly with steam or Diesel engines, it is especially important to provide positive circulation of air through the engine room. Where the ventilation is natural, a large cowl or two at the after end of the engine room introduces air, and uptakes should be provided at the forward end to carry away the heated air.

With a large boat, and especially if the yacht is to be used in warm latitudes, it is desirable to provide an electric blower to supplement natural ventilation which may be inadequate when the yacht is travelling with the wind. A small blower is not expensive to install or to operate and is a valuable piece of equipment.

Cowls, skylights and ports may be provided to give natural ventilation for all living quarters, though forced ventilation is useful in hot climates. Much care should be given to ventilating compartments below deck, the space under the cabin floor and in the extreme ends of the boat. It is possible to provide a positive circulation of air to any such space without a lot of conspicuous ventilators on deck.

Electric auxiliaries are common, even on the smallest yachts, and they effect a great saving in labor. A list of these auxiliaries includes pumps, electric lighting set, water pressure outfit, blowers, windlass, boat hoists, and power steerers in the larger yachts.

Watertight subdivision is an important aid to safety in the power cruiser. It is common practice in steel yachts of about 100 to 130 feet water line to fit four watertight bulkheads, one forward, one at each end of the engine room, and one just forward of the rudder post. Larger yachts frequently have one or two more. The more, the better, of course, from the standpoint of safety and strength. In small power cruisers, little attention is usually paid to watertight subdivision on account of cost and lack of room. It is possible, however, with a little ingenuity to make a boat nonsinkable. In addition to bulkheads, flotation can be secured by the use of tanks and by filling the waste spaces with cork or balsa wood, which is very light. It is obvious that, if the cabin floor is above the water line and the entire space beneath it is occupied with water tanks, fuel tanks, air tanks and balsa wood, there is not much room for water. If further space above the water line is also buoyant, the boat is surely nonsinkable.

Boats on medium and small sized cruisers offer a vexing problem as there is little available deck space on which to carry them, and what

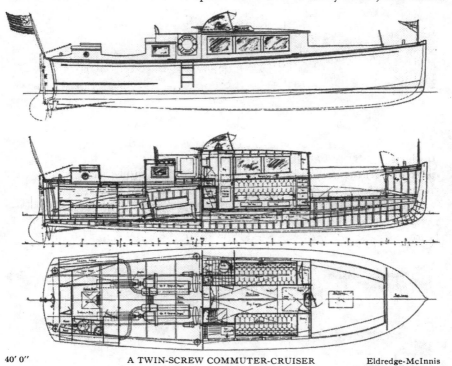

40' 0''   A TWIN-SCREW COMMUTER-CRUISER   Eldredge-McInnis

there is is apt to be rather high. Consequently the ordinary small yacht is compelled to carry a tender rather too small and too light for the best service.

The form of power yachts naturally develops along quite different lines from that of sailing yachts. Mechanical propulsion calls for length, light displacement and minimum surface, lateral plane, and stability, whereas in the sailing yacht stability or power to carry sail is a foremost consideration. The modern power yacht has a plumb stem with full deck line and generous flare to give increasing buoyancy when plunging into a head sea. This is frequently overdone, for if the

153' 6"　　　　　　　　　"ARAMIS"　　　　　　　　　Swasey

deck line is too full the vessel is slowed up perceptibly when the bow smashes deeply into a wave. The modified clipper bow as used in modern U. S. naval vessels gives a good flare without blunting the deck line too much. The trend of large motor yacht design is toward the adoption of this bow. I have used this bow in the study in Fig. 5.

Raised freeboard forward is good practice, for the gain in seaworthiness is accompanied by an increase in forecastle space. If the extra below-deck space is not needed, the main deck need not be raised and

AN OUTBOARD RACER

the high bulwarks furnish shelter for the crew and render their presence on deck less conspicuous. With a broken sheer line, a ribband at about main deck level should run continuously, in a fair line, for nearly the full length of vessel to preserve and define the true sheer.

As to sterns, the best form varies widely with the size of the boat and the speed-length ratio. In the smaller sizes, the flat or curved transom, with more or less deadrise according to speed, is most practical. In high speed cruisers, where the speed-length ratio is around 2, the transom may have considerable submergence at rest, thus straightening the lines of the bottom. (See Fig. 30.)

CHAPTER XIII

THE RACING YACHT

I N DESIGNING racing yachts we have quite a different problem from those met in other branches of yacht architecture. The purely scientific aspect is only one side of the problem and the successful designer combines theory with a large amount of experience and special skill in this branch. He studies the performance of each of his products and searches carefully for points in which the next boat may be improved. He generally thinks and talks of little else than racing boats.

A great deal of racing is done in one-design classes, especially in the smaller sizes. This type of racing educates yachtsmen but does not educate the designer much. Restricted classes are better in this respect, as in them there is more or less latitude in the shape of hull for a designer to exercise his ingenuity on. These classes ordinarily fix upon certain water line, length over all and extreme draft limits, minimum displacement, maximum sail area, and impose various scantling restrictions, as well as various other rules designed to prevent a freakish or extreme type of boat. In a class of this kind there is no chance, with all the leading characteristics fixed for him, for the designer to improve the type much.

A rating rule by which none of the factors of displacement, length or sail area is absolutely fixed, offers much greater chance for the designer to exercise his skill for the improvement of racing yachts in general. The Universal Rule is, as the name implies, widely used by the leading clubs in the United States. The introduction of this rule led to immediate and great improvement in racing yachts from the standpoint of seaworthiness and general usefulness, aside from racing.

By this rule rating measurement equals 18 per cent of the product of length, times the square root of sail area, divided by the cube root of the displacement in cubic feet. Yawls are rated at 93 per cent and schooners at 90 per cent of their measurement as determined from this formula.

L in the above formula is the length on the water line plus a penalty in the case of full bodied boats of one half the excess of the quarter-beam length over the percentage of load water line length given by the formula of $100 - \sqrt{L.W.L.}$ This L measurement is expressed mathematically as follows:

$$L = L.W.L. + .5\left[q.\ b.\ l. - \left(\frac{100 - \sqrt{L.W.L.}}{100} \times L.W.L. \right) \right]$$

It is doubtful if it ever pays to take any penalty on the quarter-beam, so that L usually equals $L.W.L.$ The quarter-beam length is measured in

30′ 0″   EIGHT-METRE SLOOP "GYPSY"   Paine

a horizontal line parallel with the middle fore and aft vertical plane at a distance from it equal to one quarter of the maximum load water line breadth and one-tenth of this breadth above the load water line plane. The hull measurements are graphically explained in Fig. 102.

In the formula we have three variables: length, sail area and displacement. Any increase in length or sail area tends to increase the

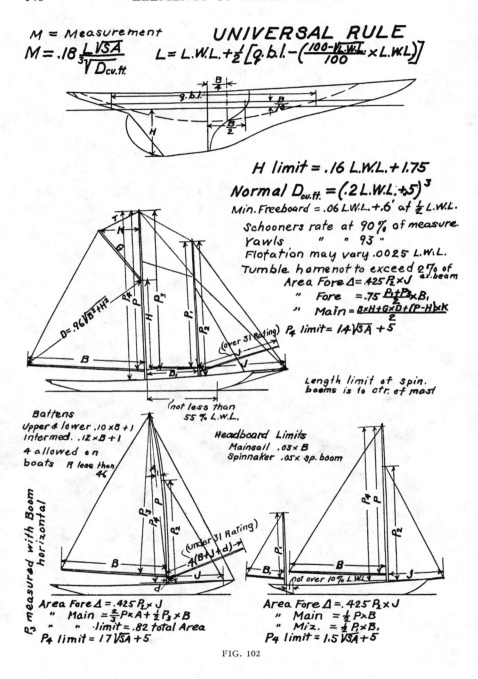

$M = Measurement$

$$M = .18 \frac{L\sqrt{SA}}{\sqrt[3]{D_{cu.ft.}}}$$

UNIVERSAL RULE

$$L = L.W.L. + \tfrac{1}{2}\left[q.b.l. - \left(\frac{100 - \sqrt{L.W.L.}}{100} \times L.W.L. \right) \right]$$

H limit $= .16$ L.W.L. $+ 1.75$

Normal $D_{cu.ft.} = (.2\ L.W.L. + 5)^3$

Min. Freeboard $= .06$ L.W.L. $+ .6'$ at $\tfrac{1}{2}$ L.W.L.

Schooners rate at 90% of measure.

Yawls   "   " 93 "

Flotation may vary .0025 L.W.L.

Tumble home not to exceed 2% of ext. beam

Area Fore $\triangle = .425\ P_2 \times J$

"   Fore $= .75 \frac{P_1 + P_3}{2} \times B_1$

"   Main $= \frac{B \times H + G \times D + (P - H) \times K}{2}$

P_4 limit $= 1.4\sqrt{SA} + 5$

$D = .96\sqrt{B^2 + H^2}$

(over 31 Rating)

Length limit of spin. booms is to ctr. of mast

(not less than 55% L.W.L.

Battens

Upper & lower $.10 \times B + 1$

Intermed. $.12 \times B + 1$

4 allowed on boats R less than 46

Headboard Limits

Mainsail $.03 \times B$

Spinnaker $.05 \times$ sp. boom

P_3 measured with Boom horizontal

(under 31 Rating)

$.1(B + J + d)$

Area Fore $\triangle = .425\ P_2 \times J$

"   Main $= \tfrac{2}{3} P \times A + \tfrac{1}{2} P_3 \times B$

"   "   limit $= .82$ total Area

P_4 limit $= 17\sqrt{SA} + 5$

not over 10% L.W.L.

Area Fore $\triangle = .425\ P_2 \times J$

"   Main $= \tfrac{1}{2} P \times B$

"   Miz. $= \tfrac{1}{2} P \times B_1$

P_4 limit $= 1.5\sqrt{SA} + 5$

FIG. 102

rating since they are in the numerator and any increase in displacement up to a certain limit tends to decrease the rating. Actually, length
divided by the cube root of the displacement remains practically constant with the result that the sail area is practically fixed for a given
rating, decreasing only very slowly as the length increases.

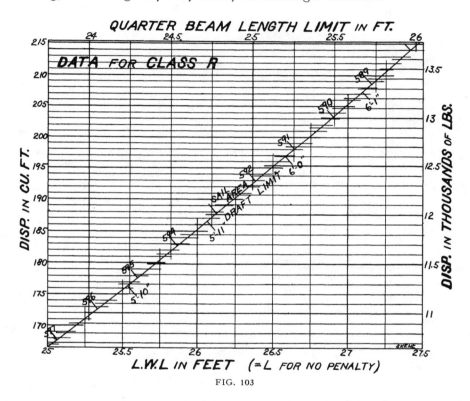

FIG. 103

Sail area in the formula is not actual area but is figured in an arbitrary manner for fore triangle and the other sails, as indicated in Fig.
102. Area is given in square feet.

Displacement, which is taken in cubic feet in the formula, is controlled directly by the length owing to the limitation that the cube
root of displacement may not be taken in the formula at greater than
20 per cent of the L.W.L. plus .5, even though it may actually be
greater than that. The natural course is to take the displacement as
provided by this formula. If taken at less than this amount, the sail
area will have to be reduced accordingly; and if taken at greater than
this amount, the resistance will be increased without any additional

sail area being permissible to offset it. This limiting displacement I will term the normal displacement.

Fig. 23 gives normal displacements up to 30 ft. *L.W.L.* Fig. 24 gives normal displacements for yachts 30 to 60 ft. *L.W.L.* and Fig. 25 for yachts over 60 ft. *L.W.L.* On all these curves the displacement is indicated in cubic feet as well as by weight and the draft limit is also indi-

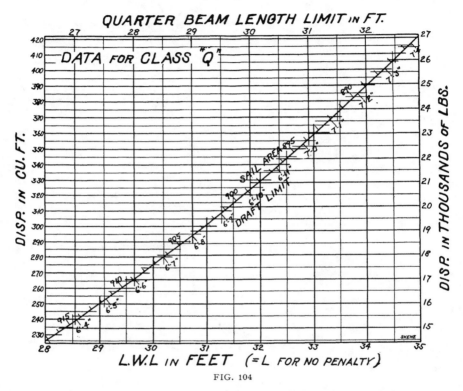

FIG. 104

cated. Fig. 103 gives the principal characteristics of the "R" Class. Fig. 104 does the same for Class "Q".

The Universal Rule is really a sail area rule, that is, for every different rating there is an amount of sail allowed that varies but little within the range of practicable water line lengths.

Take the rule $R=\dfrac{.18 \times L \times \sqrt{S\,A}}{\sqrt[3]{D}}$ and substitute for $\sqrt[3]{D}$ its value in

terms of L. Then $R=\dfrac{.18 \times L}{.2\,L+.5} \times \sqrt{S\,A}$.

Replacing the fraction $\dfrac{.18\,L}{.2\,L+.5}$ by a single character K, we have the

simple rule $R = K\sqrt{S\,A}$ from which $S\,A = \left(\dfrac{R}{K}\right)^2$.

44' 6"   TWELVE-METRE CLASS "SEVEN SEAS"   Crane

Values of K, which of course vary somewhat with the length, are given in Fig. 105. Fig. 106 shows the limit sail area for the three small rating classes, S, R and Q, and Fig. 107 gives it on a less accurate scale for all the classes, between limits where the length is $.9\,L$ and $.7\,L$.

On general principles, size is an advantage, especially where it may be accompanied by increase in sail area, which, however, is not possible under this rule. It certainly pays, however, to make the length as

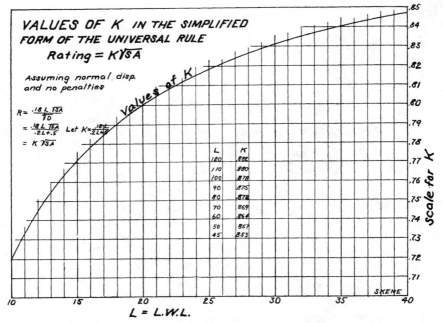

FIG. 105

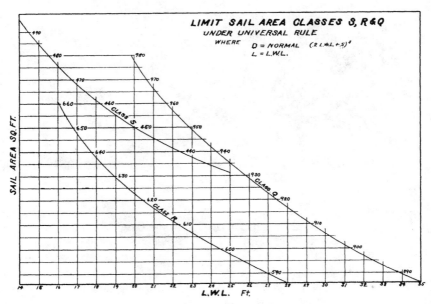

FIG. 106

great as possible without having the resulting sail area too small to
drive the boat under average conditions. Shorter and fuller boats have
been tried with the idea that the proportionally larger sail area and
lighter displacement can offset the shorter length, which has not
proved to be the case. In almost any moderate-speed vessel, the principal factor for speed is length, and it is possible that even longer boats
than have been used in the smaller rating classes will prove to be
superior. This is doubtful, however, for limit in size appears to have
been reached. The increase in length beyond precedent in a given

30′ 0″   ONE OF THE FAMOUS NEW YORK "30s"   Herreshoff

class and the additional displacement involved, would be distributed
below the bilges where it would cause the least increase in resistance.
The increased ballast which accompanies the increase in displacement
increases the stability due to weight so that it is desirable to decrease
rather than increase the stability of form. Extreme development along
these lines approaches the narrow plank-on-edge and is limited by the
restriction that the L.W.L. must not exceed 1.08 times the measured
length + .5. Our ideas of racing form required considerable readjust-
ment with the introduction of the Universal Rule and this readjust-
ment is not yet completed. The racing yacht of a few years hence is
likely to show considerable change from present-day models.

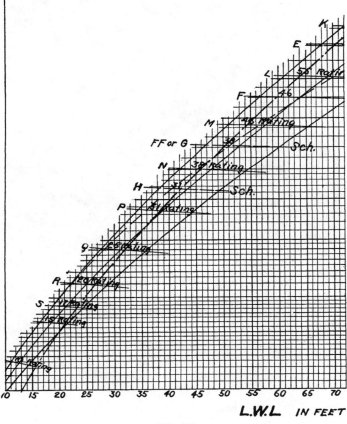

FIG. 107

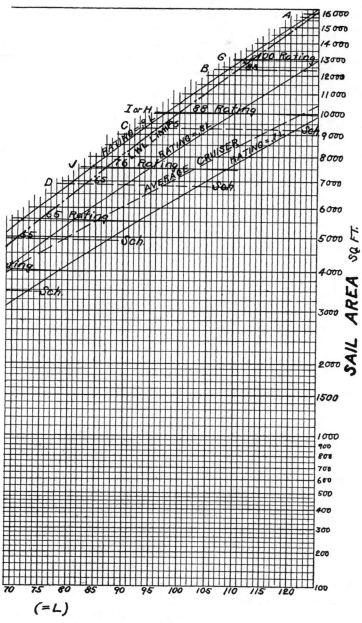

FIG. 107

The encouragement which this rule places on length produces a much narrower and smaller rigged boat than the normal cruising yacht for a given length. A fairer basis of comparison between cruisers and racers than length, however, is sail area, just as horse power is the best basis for comparing mechanically driven boats or vehicles. On the amount of sail carried depends in a general way the size of crew necessary. On this basis, comparing out and out cruisers with Universal Rule boats of the same sail area, it will be found that freeboard, displacement and internal space are about the same in both —

FIG. 108

that the cruiser is shorter and wider, with less ballast in proportion to displacement. These differences are much less in the larger sizes.

Sail plan of an R class yacht is shown in Fig. 108 and her construction plan in Fig. 110.

THE INTERNATIONAL RULE OF MEASUREMENT

Racing abroad is carried on largely under the International Racing Rule which is set forth and explained in Fig. 111. This is a more complicated rule than the Universal, having five variables. Displacement does not figure directly in the rule, but is really specified by the stipulation that the cube root of the displacement should be not less than .2

L.W.L. plus .5. The minimum beam, measured at the point of extreme beam, at one-third of the rule midship freeboard above the water

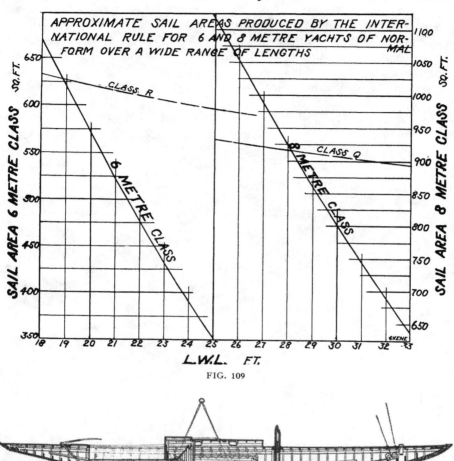

FIG. 109

FIG. 110

line shall be: For a Six-Metre, 6 feet; Eight-Metre, 8 feet; Ten-Metre, 9.9 feet; Twelve-Metre, 11.8 feet. This limitation applies to all boats built after September 22, 1937. As to sail area, the area of the fore triangle is computed exactly as under the Universal Rule but the actual area of the mainsail is taken. It will be noted that freeboard is encouraged since this is subtracted from the numerator. A series of sail areas corresponding to given lengths cannot be computed since there are so many variables in the formula but for boats of average proportions the sail areas would run about as shown in Fig. 109 for Six-Metre and Eight-Metre yachts, showing the tremendous variation in sail area possible under this rule.

A fine-lined yacht is compelled by the tax on girth at forward and

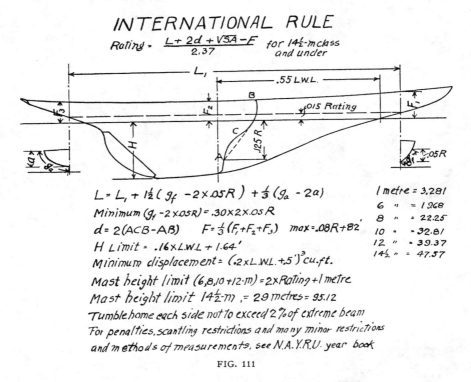

INTERNATIONAL RULE

Rating = $\dfrac{L + 2d + \sqrt{SA} - F}{2.37}$ for $14\frac{1}{2}$-m class and under

$L = L_1 + 1\frac{1}{2}(g_f - 2 \times .05R) + \frac{1}{3}(g_a - 2a)$

Minimum $(g_f - 2 \times .05R) = .30 \times 2 \times .05R$

$d = 2(ACB - AB)$ $F = \frac{1}{3}(F_1 + F_2 + F_3)$ max $= .08R + 82$

H Limit $= .16 \times L.W.L + 1.64'$

Minimum displacement $= (.2 \times L.W.L + .5')^3$ cu.ft.

Mast height limit $(6,8,10 + 12\text{-}m) = 2 \times Rating + 1$ metre

Mast height limit $14\frac{1}{2}\text{-}m$, $= 29$ metres $= 95.12$

Tumblehome each side not to exceed 2% of extreme beam

For penalties, scantling restrictions and many minor restrictions and methods of measurements, see N.A.Y.R.U. year book

1 metre	= 3.281
6 "	= 19.68
8 "	= 22.25
10 "	= 32.81
12 "	= 39.37
14½ "	= 47.57

FIG. 111

after points of measurement. Under the Universal Rule, boats of the old scow types have proved failures on account of the short water line necessitated through the girth tax.

Both these rules, while more or less imperfect, are great steps in advance and boats produced by them are able, seaworthy craft, suitable for pleasure purposes aside from racing. Scantling restrictions, which will be touched upon in another chapter, are imposed which insure substantial construction.

The displacements required with both these rules are sufficiently large so that a liberal amount of ballast is needed to give the proper flotation. In some boats this amounts to as much as 75 per cent of the displacement. With stability resulting so largely from a low center of

RULES AND LIMITING DIMENSIONS FOR SCHÄREN-KREUZERS

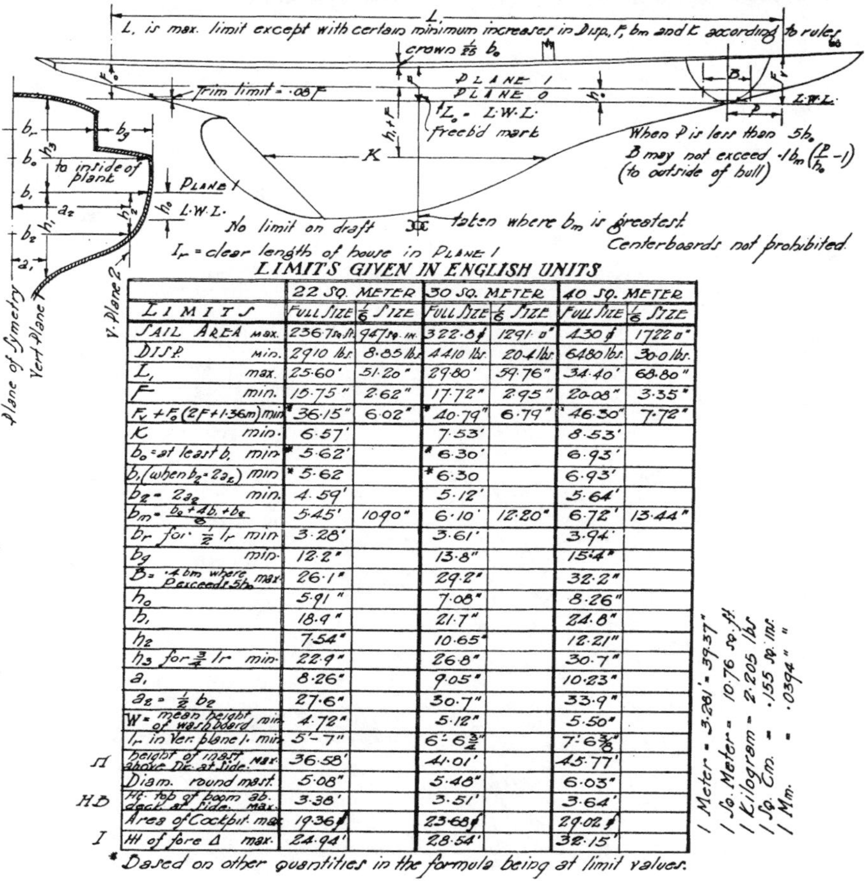

L, is max. limit except with certain minimum increases in Disp, F, bm and K according to rules

Trim limit = .08 F
PLANE 1
PLANE 0
L. = L.W.L.
free'b'd mark

When P is less than 5b₀
B may not exceed .1bm (P/5b₀ −1)
(to outside of hull)

No limit on draft. III taken where bm is greatest.
Centerboards not prohibited
l_r = clear length of house in PLANE 1

LIMITS GIVEN IN ENGLISH UNITS

LIMITS	22 SQ. METER		30 SQ. METER		40 SQ. METER	
	FULL SIZE	1/6 SIZE	FULL SIZE	1/6 SIZE	FULL SIZE	1/6 SIZE
SAIL AREA max.	236.7 sq.ft	9.57 sq.in	322.3	1291.0	430.0	1722.0
DISP. min.	2910 lbs	8.85 lbs	4410 lbs	20.4 lbs	6480 lbs	30.0 lbs
L max.	25.60'	51.20"	29.80'	59.76"	34.40'	68.80"
F min.	15.75"	2.62"	17.72"	2.95"	20.08"	3.35"
F_0+F_0 (2F+1.36m) min	36.15"	6.02"	40.79"	6.79"	46.30"	7.72"
K min.	6.57'		7.53'		8.53'	
b_0 at least b_i min	*5.62'		*6.30'		6.93'	
b_i (when $b_2 = 2a_2$) min	*5.62		*6.30		6.93'	
$b_2 = 2a_2$ min.	4.59'		5.12'		5.64'	
$b_m = \frac{b_0+4b_i+b_2}{6}$	5.45'	10.90	6.10'	12.20	6.72'	13.44"
b_r for $\frac{1}{2} l_r$ min	3.28'		3.61'		3.94'	
b_3 min	12.2"		13.8"		15.4"	
B = .4 bm where P exceeds 5b₀ max	26.1"		29.2"		32.2"	
h_0	5.91"		7.08"		8.26"	
h_i	18.9"		21.7"		24.6"	
h_2	7.54"		10.65"		12.21"	
h_3 for $\frac{3}{4} l_r$ min	22.9"		26.8"		30.7"	
a_i	8.26"		9.05"		10.23"	
$a_2 = \frac{1}{2} b_2$	27.6"		30.7"		33.9"	
W = mean height of washboard min	4.72"		5.12"		5.50"	
l_r in Ver. plane l. min	5'-7"		6'-6 3/4"		7'-6 5/8"	
height of mast above DK at side max	36.58'		41.01'		45.77'	
Diam. round mast	5.08"		5.48"		6.03"	
Ht. Top of boom ab. deck at side max	3.38'		3.51'		3.64'	
Area of Cockpit max	19.36		23.68		29.02	
Ht of fore △ max	24.94'		28.54'		32.15'	

*Based on other quantities in the formula being at limit values.

1 Meter = 3.281 sq.ft = 39.37"
1 Sq.Meter = 10.76 sq.ft.
1 Kilogram = 2.205 lbs
1 Sq.Cm. = .155 sq.ins.
1 Mm. = .0394"

FIG. 112

gravity, the boats are non-capsizable and it is seldom necessary to reef.

A foreign racing type which is becoming popular in America is the Schären-Kreuzer. This is a highly developed light displacement type where sail area is the principal restriction though there are numerous others as set forth in Fig. 112. Comparing this type with a Universal Rule boat, for a given L.W.L. the sail area and displacement are roughly twice as great for the latter. A typical profile of a 22 Square Metre boat is shown in Fig. 113. The method of measuring sail is shown in Fig. 114. This type may be driven at high speeds in strong winds on account of the light displacement in proportion to length.

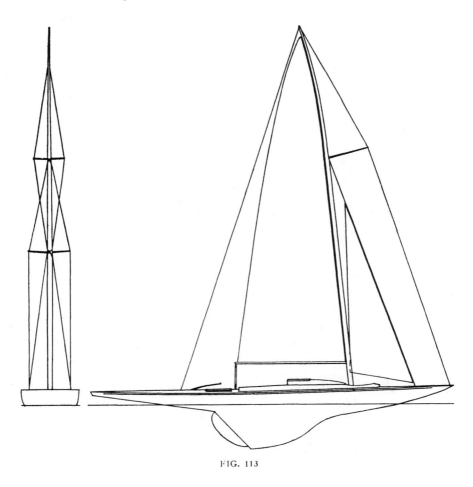

FIG. 113

Fig. 115 shows the Bermuda Race Rule, that of the Cruising Club of America for ocean racing. It was adopted by the Governing Board on November 4th, 1937, to be effective for three years from date, and is as follows:

RATING. $.95 \ (L \pm B \pm D \pm P \pm S \pm F - I + A + C) \times$ Propeller Factor $\times R$.

"L." For the purpose of this rule, a yacht's length shall be the sum of 30 per cent of the L.W.L. and 70 per cent of the length at 4 per cent Water Line Plane.

L = .3 L.W.L. + .7 Length at 4 per cent Water Line Plane.

LENGTH AT 4 PER CENT WATER LINE. The length of a water line in a plane 4 per cent of the L.W.L. above the L.W.L., corrected for jogs, notches or hollows in the profile, as follows:

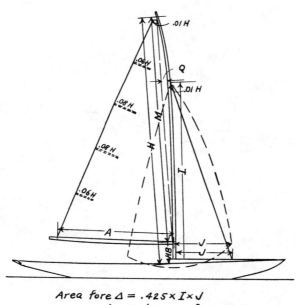

Area fore △ = .425 × I × J
"   mains'l = ½ A × M + ⅔ Q × M

FIG. 114

The points of intersection of this plane and the profile must extend beyond the L.W.L. endings.

Any local concave jog or notch (curved or angular) at the plane of measurement, at either end, shall be bridged by a straight line and the 4 per cent length taken to the intersection of such lines with the established 4 per cent W.L. plane. Except that a concave bow profile, producing what is commonly known as a clipper bow, shall be permitted without bridging, provided that all the lines of such bow clearly indicate that it is a true clipper bow, and not hollowed in profile for the purpose of unfairly reducing the 4 per cent length measurement.

"L.W.L." L.W.L. shall be the length of the load water line determined by subtracting from the overall length the forward and after overhangs. Overall length shall be the length from the aftermost part of the counter or taffrail to the intersection of the forward side of the stem and the top of the covering board, or the extension of either, or both if necessary. When *L.W.L.* is measured, all ballast and movable gear which is to be carried below the cabin floor while racing shall be so placed, and no ballast or movable gear not so placed at the time of measurement may be carried below the floor boards during any race, nor may any inside ballast or movable gear stored below the floor boards at the time of measurement be carried other than in its loca-

BERMUDA RACE RULE

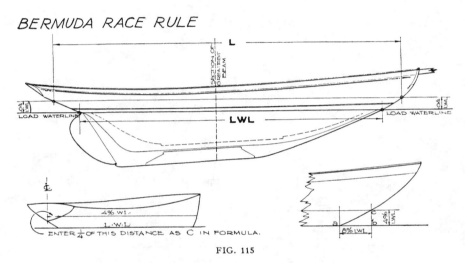

FIG. 115

tion at the time of measurement. It is not necessary that food, ice, water and other cruising stores be aboard at the time of measurement.

"*B.*" For the purpose of this rule, a yacht's beam shall be the average of her beams at the load water line and the 4 per cent water line, taken at the point of maximum beam at the load water line.

(a) If a yacht's beam be less than .187 *L* plus 3.2, the difference between her beam and .187 *L* plus 3.2 shall be multiplied by 2 and inserted in the formula as a plus quantity.

(b) If a yacht's beam exceeds .187 *L* plus 3.2 but is less than .21 *L* plus 3.8, the difference between her beam and .187 *L* plus 3.2 shall be multiplied by 1.25 and inserted in the formula as a minus quantity.

(c) If a yacht's beam exceeds .21 *L* plus 3.8, multiply the difference between .21 *L* plus 3.8 and .187 *L* plus 3.2 by 1.25 and add to that amount one-third of the excess over .21 L plus 3.8 and insert in the formula as a minus quantity.

"*D*." A yacht's base draft shall be .147 L plus 1.5, above which limits penalties shall be assessed and below which credits shall be given as follows:

25' 3"   "GYPSY"   Paine

If the yacht's draft exceeds the base draft, the difference shall be multiplied by 1.5 and inserted in the formula as a plus quantity.

If the yacht's draft be less than base, the difference shall be multiplied by .75, in the case of keel yachts, and inserted in the formula as a minus quantity, and, in the case of centerboard yachts, such difference shall be multiplied by .1875 and inserted in the formula as a minus quantity.

"*I.*" Yachts having iron keels shall receive an additional credit of .0185 *L.*

"*P.*" The cube root of a yacht's Base Displacement (referred to in

23' 5"   SIX-METRE CLASS "CHALLENGE"   Luders

the following as "cube root *B.D.*") shall be .179 *L* plus .8, below which limit a penalty shall be assessed, and above which credit shall be given as follows:

  (a) If the cube root of a yacht's displacement (in cubic feet) is less than cube root *B.D.*, the difference between the cube root of her measured displacement and cube root *B.D.* shall be multiplied by 4.5 and inserted in the formula as a plus quantity.

(b) If the cube root of a yacht's displacement (in cubic feet) exceeds cube root *B.D.* by not more than 10 per cent, the difference between the cube root of her displacement and cube root *B.D.*

26' 0"   R CLASS SLOOP "MOJOLA"   L. F. Herreshoff

shall be multiplied by 3.5 and inserted in the formula as a minus quantity.

(c) If the cube root of a yacht's displacement (in cubic feet) exceeds cube root *B.D.* by more than 10 per cent, the value of *P* equals .35 cube root of *B.D.* plus twice the difference between 1.1 cube root *B.D.* and cube root of yacht's displacement.

"*S.*" The square root of a yacht's Base Sail Area shall be four times the cube root of the base displacement ($4\sqrt[3]{BD}$), below which limit a credit shall be given and above which a penalty shall be assessed as follows:

(a) $S = \sqrt{SA} \times$ Rig Allowance $- 4\sqrt[3]{BD} = $ a plus quantity in the formula.

If the square root of sail area times rig allowance is less than 4 times cube root of *B.D.*, multiply the difference by .75 and insert in the formula as a minus quantity.

(b) $S = 4\sqrt[3]{BD} - \sqrt{SA} \times$ Rig Allowance $=$ a minus quantity in the formula.

In this formula *SA* (Sail Area), to be measured as by the N.A.Y.R.U. rules as adopted and published with the following changes:

In the event that the area of the mizzen of a yawl be less than 10 per cent, or that of a ketch less than 18 per cent of the total sail area as measured without credits or penalties, an additional amount of sail area shall be added to the formula to bring the area of the mizzen up to these percentages.

Mechanically bent booms shall be barred.

MAST HEIGHT DEBITS AND CREDITS. The following shall be the base height of masts for jib-headed mainsails, above which heights penalties will be assessed and below which credits will be given — measurements to be taken from the deck (or in the case of cabin yachts, from the deck projected) to the highest point of measurement:

Single-masted yachts. $1.7\sqrt{SA}$ plus 5 ft.
Ketches and yawls. $1.7\sqrt{SA - \text{Area of Mizzen}}$, plus 5 ft.
Schooners. $1.7\sqrt{SA} - .85$ Fore Triangle plus 5 ft.

Any excess height above these limits to be multiplied by 3 and added to the perpendicular *P* in computing the area of the mainsail. Any deficiency below these limits to be divided by 2 and subtracted from the perpendicular *P* in computing the area of the mainsail.

RIG ALLOWANCES. Values for rig allowances to be varied depending on the course to be sailed and the type of weather anticipated. The rig allowances for the 1938 Bermuda Race shall be as follows:

Jib-headed sloops and cutters. 1.00
Jib-headed yawls. 98
Gaff-headed sloops and cutters, staysail ketches. 97
Jib-headed staysail schooners. 96

Gaff-headed yawls, schooners with jib-headed main and gaff
foresail... .93
Jib-headed ketches....................................... .92
Gaff-headed schooners and gaff-headed ketches............ .90

"*F.*" To be the average of the freeboard taken to the top of the
covering board at the bow and stern endings of the *L.W.L.* plus one-
half the average of the rail height, including the rail cap at these
points. The average of rail heights not to exceed 8″ for credit.

(a) If the freeboard as measured above be less than .0566 *L* plus
1.1 ft., the difference is to be multiplied by 2 and inserted in the
formula as a plus quantity.

(b) If the freeboard be in excess of .0566 *L* plus 1.1 but less than
.069 *L* plus 1.2, the difference is to be multiplied by 1.5 and
inserted in the formula as a minus quantity.

(c) If freeboard exceeds .069 *L* plus 1.2, multiply the difference
between .0566 *L* plus 1.1 and .069 *L* plus 1.2 by 1.5 and add
to that amount .75 times the excess over .069 *L* plus 1.2 and
insert in the formula as a minus quantity.

"*A.*" If the horizontal distance between the forward end of the
4 per cent water line and the forward end of the *L.W.L.* exceeds 8
per cent of the *L.W.L.* length (twice the height of the 4 per cent plane
above *L.W.L.* plane), the excess is to be entered in the formula as a
plus quantity.

"*C.*" Should the 4 per cent plane intersect the stern transom, one-
quarter of the distance from the edge of the transom, at the point of
intersection, to the center line, shall be entered in the formula as a plus
quantity.

PROPELLER FACTORS.

Folding propeller, on center but not in deadwood.......... 1.00
Feathering propellers on center but not in deadwood....... .99
Folding propeller, off center............................ .98
Feathering propeller, off center, or folding propeller in dead-
wood... .97
Solid propeller, off center.............................. .96
Feathering propeller, in deadwood........................ .95
Solid propeller, in deadwood............................. .95

"*R.*" BALLAST-DISPLACEMENT RATIO. The Ballast-Displacement
ratio shall be the ratio of the weight of total ballast to displacement in
pounds. In computing the total ballast, its weight shall be the sum of

the weights of the fixed, or outside ballast, plus the inside ballast, plus the weight of any movable gear stored below the cabin floor. Ballast-Displacement ratio will hereafter be referred to as B/D ratio.

The base B/D shall be .43. Above this ratio ratings shall be increased and below it they shall be decreased by an application of the R factors of the following schedule:

B/D Ratio	"R" Factor
.451 to .46	1.030
.441 " .45	1.016
.431 " .44	1.007
.421 " .43	1.000
.411 " .42	.994
.401 " .41	.989
.391 " .40	.984
.381 " .39	.979
.371 " .38	.975
.361 " .37	.972
.351 " .36	.969
.341 " .35	.967
.331 " .34	.965
.321 " .33	.963
.311 " .32	.961
Under .311	.960

$L = L.B.G.$ plus or minus freeboard correction.

$L.B.G.$ is the distance between the forward and after girth stations. The forward girth station ($F.G.S.$) is the position forward at which the girth taken vertically from covering board to covering board is equal to half B. The after girth station ($A.G.S.$) is the position aft where the girth measurement from covering board to covering board is equal to three-quarters B.

"Covering board" shall be taken as the point where the extension of the curve of the top of the deck intersects with the curve of the side.

Freeboard Correction: If the sum of the freeboards taken at $F.G.S.$ and $A.G.S.$ is less than 10 per cent of $L.B.G.$ plus 2.5 feet, then twice such deficit shall be added to $L.B.G.;$ or if it is greater, any excess up to 4 per cent of $L.B.G.$ shall be deducted from $L.B.G.$

$B.$ The fore and aft position of the greatest beam must be found. B shall be the greatest breadth, in this fore and aft position, measured to the outside of normal planking at a height not exceeding half the freeboard height.

$D.$ The formula shall be:

$$D = \frac{1.5\,FD + MD}{2} + .2 \text{ height of bulwarks.}$$

Depth shall be taken in two places, forward (FD) at 25 per cent of $L.B.G.$ abaft $F.G.S.$, and amidships (MD) at 50 per cent of $L.B.G.$ abaft $F.G.S.$ Both measurements shall be taken vertically from a straight line joining the under side of the deck at sides of vessel; FD to a point one-tenth of B out from the center line, on the inside of wood planking; MD to the top of the keel on the center line.

MD shall not be taken as exceeding the maximum draft limit.

The MD measurement shall not be taken as exceeding one and a half times "quarter beam depth." "Quarter beam depth" is taken vertically from the same height in the same fore and aft position, to a point on the inside of wood planking one-quarter of B out from the center line.

Bulwark height shall be measured vertically from the top of rail to deck amidships.

NOTE: When taking these measurements in steel vessels, allowance to be made for molding of wood keel and thickness of planking.

Sheer. If the sheer is not a continuous concave curve, three times any excess of freeboard at MD over half the sum of the freeboards at $F.G.S.$ and $A.G.S.$ shall be subtracted from D.

Draft. When the vessel is in seagoing trim, draft, including centerboards if fitted, shall not without penalty exceed 16 per cent of the length of water line plus 2 feet. One and a half times any excess shall be added to the measured rating.

$\sqrt{S}$. The square root of sail area ($S.A.$) after any penalities have been applied, multiplied by rig allowance ($R.A.$).

Rig Allowance is expressed as a percentage of the square root of $S.A.$

Jib-headed cutter	100
Jib-headed yawl	98
Jib-headed or wishbone schooner and gaff cutter	96
Jib-headed or wishbone ketch and gaff yawl	94
Gaff schooner	92
Gaff ketch	90

NOTE: The denomination of the rig, whether jib-headed or gaff, shall be determined by the type of the mainsail.

There follow rules for measuring the sail area and for the limiting sizes of spinnakers, headsails, mast heights, etc., as well as scantlings and propeller allowance. For these, the reader is referred to the year book of the Royal Ocean Racing Club.

CHAPTER XIV

CONSTRUCTION

A<small>N INTIMATE</small> acquaintance with methods of construction is imperative to the successful designer, as in many cases the form must be adapted to the method of construction used in building the yacht. This is particularly true of the region of the fin in sailing yachts. Moreover, the designer must be thoroughly familiar with the method of construction to be used in building, that he may determine what the weights will be and provide sufficient displacement.

The hull of a yacht, particularly an extreme racing yacht, is subjected to a complicated system of stresses, among which are those produced by the forces acting on the mast, by the leverage of keel or centerboard, by the impact of waves against the hull, and by the tension of the rigging. The designer must thoroughly understand the nature of these stresses to be able to design an efficient structure to withstand them. By an efficient structure is meant one in which there is no useless weight, each member being carefully proportioned to the work it has to do. As displacement is the principal factor for resistance, the saving of weight is of the highest importance where extreme speed is sought. To save weight systematically, each member entering into the structure of the yacht must be carefully designed for the highest efficiency. This necessitates a thorough knowledge of the physical properties of the various materials used in the construction.

The outer skin is the principal member of the yacht's structure as it is the realization of the desired form and is, in a sense, the boat herself. All other members may be regarded as auxiliaries in assisting the skin to perform its function. The principal member for longitudinal strength is the keel, though the skin, deck, stringers, etc., contribute materially. The transverse strength is supplied mainly by the frames and deck beams. These are assisted in their work by stringers, clamps, floors and knees.

There are various methods of determining suitable scantlings or sizes of the various members of the yacht's structure. The most natural way is to be guided by the scantlings of an existing yacht of about the desired size and type which has proved herself under service conditions to be strong enough and structurally well proportioned. Racing boats, as a rule, have their scantlings specified by the rules of the particular class for which they are built. It is theoretically possible to compute the necessary sizes for the principal members, but this method is utterly impracticable for ordinary work on account of the immense amount of mathematical labor involved as well as the uncertainty of the exact nature of the stresses to which the yacht is subjected. Something along this line may be done, however, in the case of large yachts of extreme or unusual proportions. In the case of large cruising yachts, the building rules of one of the classification societies furnish an excellent guide for the determination of suitable scantlings. These building rules are based upon the result of practice as well as upon the deductions of scientific research, and may be relied on to produce a substantial and durable hull.

CLASSIFICATION SOCIETIES

The classification societies whose rules are most used for yacht construction in this country are Lloyd's Register of Shipping and the American Bureau of Shipping. Lloyd's Register publishes a set of rules especially drawn up for yacht construction in steel, composite and wood. These rules are applicable to yachts of all sizes from about 20 feet water line upwards. The scantlings required under these rules are somewhat in excess of those usual to American practice, particularly in the smaller sizes. Under Lloyd's rules the principal scantlings are assigned according to what are known as transverse and longitudinal numbers, which are based on length, beam and depth.

The transverse number regulates the sizes of frames and floors. The longitudinal number regulates the dimensions of planking, keel, stem, sternpost, rudder stock, shelves and fastenings. The beams are proportioned according to their length.

STEEL CONSTRUCTION

There are three general forms of yacht construction, the all metal, the all wood and the composite. These will be dealt with briefly in order. The all metal construction is the most common for yachts of over 80 feet water line. The advantages of this construction are great strength, durability and increased cabin accommodations. Steel is the usual material used, although bronze is sometimes employed in large racing yachts. It is doubtful whether the smoother surface of bronze compensates for its greater weight as compared with steel.

The midship section of the 100-foot schooner yacht shown in Fig. 116 will serve to illustrate the principal features of construction in steel. The sizes of members are expressed as follows: plates in pounds per square foot of area or in twentieths of an inch thickness. A steel plate one-inch thick weighs about 40 pounds per square foot or two pounds for each twentieth of an inch thickness. Angle bars, I bars, channel bars, etc., are rated according to the dimensions of their legs, and their weight per lineal foot or thickness in twentieths of an inch.

The keel of the schooner shown in Fig. 116 is a 25-pound plate flanged at the sides and riveted to the garboard strake. The latter and the sheer strake are heavier than the remainder of the plating, as

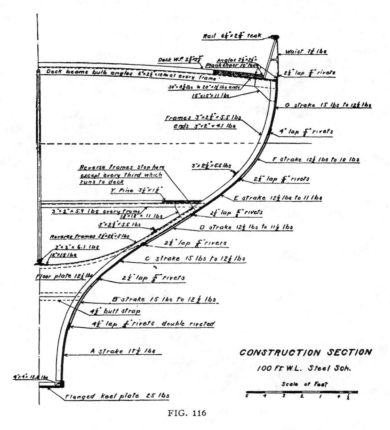

FIG. 116

they are subjected to greater stresses. The system of plating shown is the ordinary "clencher" or "in and out" style. Flush plating is often used above the water line on yachts for the sake of appearance, but it is heavier, more expensive, and not so strong as the "clencher" system. The frames are steel angles $3'' \times 2\frac{1}{2}'' \times 5.5$ pounds, spaced 22''. Each pair of frames is tied together at the feet by floor plates. The keel plate is attached to the floor plate by short angle clips. Deep frames, consisting of plate riveted to ordinary frames, are fitted at the masts to

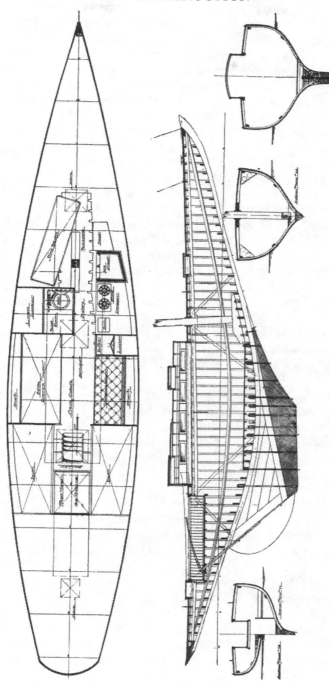

CLASS Q BOAT DESIGNED BY FREDERICK M. HOYT

take the strain of the rigging. Angle irons, known as reverse frames, extend across the top of each floor plate and are riveted to the backs of the frames. These tie the frames together and stiffen the floor plate. The keelson shown consists of a flat plate and two angles riveted together. The deck beams are bulb angles and are supported in the middle by steel pillars. They are attached to the frames by bracket plates. The brackets are often welded on the beam itself.

A deck stringer plate is worked all around the vessel between the beams and the wood deck. It is connected to the sheer strake by an angle bar as shown. Another angle, some ten inches inboard and parallel to the deck line, is riveted to the stringer plate, and the space between the two angles forms a waterway to drain water to the scuppers. The bulwarks are of 7½-pound plating, riveted to the sheer strake and supported at frequent intervals by stanchions. They should be capped by a broad teak rail. The deck is of white pine plank laid parallel to the center line and bolted to the deck beams. The bilge stringer consists of a pair of angles riveted back to back. Steel plates are fitted at each mast beneath the wood deck, and these are connected to the deck stringer plate by diagonal tie plates to take the strain of masts and prevent the deck from wringing.

The ballast is lead stowed inside the fin and covered with a layer of cement. Limber holes are cut through the floor plates at the surface of the cement to allow bilge water to drain to the lowest point.

WOOD CONSTRUCTION

The all wood construction is almost universal in this country for yachts of less than 50 feet water line. Fig. 117 is the construction plan of the 30-footer and may be taken as typical of construction in wood. The following paragraphs are descriptive of the various members and of the customary materials, with their relative values.

Planking — The woods commonly used for planking of yachts are yellow or hard pine, white cedar, white pine, Spanish cedar and mahogany. Other woods, such as cypress and oak, are occasionally used, but they are less suitable than those first mentioned. Yellow pine is a tough, durable wood and may be obtained in long lengths. It is quite heavy and for that reason its use is confined to cruising boats. White cedar is a light, easily-worked wood and does not absorb water badly. It is obtainable in short lengths only, which necessitates a large number of butts in a boat of any length. White pine much resembles cedar, but is less durable. It is used largely for the inner skin of double-planked boats. Spanish cedar and mahogany are quite extensively used for planking, especially on racing boats, although they are con-

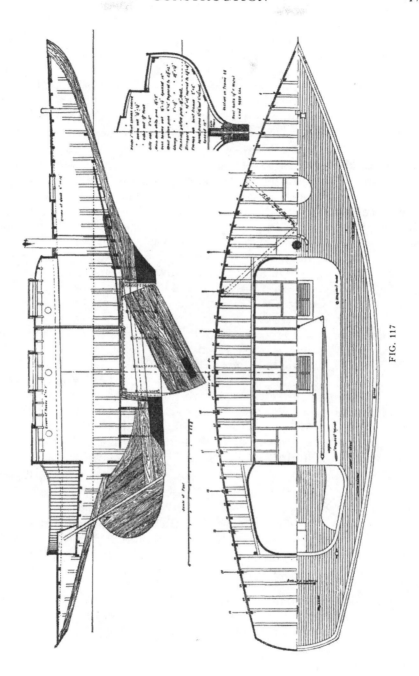

FIG. 117

siderably heavier than cedar. Teak is heavy, but is the most durable wood known and is unsurpassed for planking where its weight and cost are not prohibitive. It holds fastenings excellently and does not come and go with the weather. The use of these woods is partly on account of their handsome appearance when finished bright, but more especially because they stay in place and retain a smooth finish better than other woods. Double planking is quite extensively used on expensively built boats. The advantages claimed for double planking are greater strength and less liability of leaking, less weight, and a smoother surface. The best method of double planking is to make the inner skin about $\frac{1}{3}$ as thick as the outer, neither skin being caulked. The two skins are bonded by applying thick shellac, white lead, marine glue or casein glue thickly between the skins. This acts as a cement, binding the two skins together. Caulking is unnecessary if the outer skin is carefully fitted; it would prevent the surface being as smooth as it otherwise might be. In fastening double planking, the inner skin is first tacked in place, and then both skins are through fastened to the frames. Between the frames the inner skin should be fastened to the outer with bronze screws. The principal types of fastenings used for planking are copper nails riveted on burrs, bronze screws, chisel-point galvanized boat nails, and galvanized cut nails. Their value as fastenings stands about in the order given.

Frames — Oak is the usual material for frames, although American elm and hackmatack are admirably adapted to the purpose. Frames are of two kinds, steam bent and sawn. Steam bent frames are lighter for the same strength and are used in yachts up to about 30 feet water line. Where bent frames are used, the form of the boat is determined by moulds made from the design enlarged to full size in the mould loft. These are erected in their proper places on the keel, being spaced from two to four feet apart. After a sufficient number of longitudinal ribbands have been fastened to the moulds, the frames may be put in hot, directly from the steam box, fastening them against the ribbands where they cool in shape. The bevel is obtained by putting a twist in the frame. Some builders bend their frames around moulds to a little greater curvature than they will have in the boat; after they have dried, they are cut to the proper bevel and put into the boat cold. This method entails more labor than the first. In small craft with little deadrise, the frames may be made continuous from rail to rail. Where the frames are separate on each side, as is generally the case, the strength at the center line must be preserved by connections known as floors. Frames larger than two inches square cannot be bent readily.

Where frames are too large to bend, the sawn frame construction is used. Each frame is usually double and is built up of several short lengths from natural crooks sawn to shape. These short lengths "break joints" on the two halves of the frame, thus preserving the strength. The usual practice is to use all sawn frames in yachts over 40 feet. In yachts between 25 and 40 feet water line, a combination of sawn and bent frames is commonly used, every other or every third frame being sawn. This system of framing is used on the 30-footer (see Fig. 117). With this method the boat is ribbanded up on the sawn frames, the bent frames being put in in the usual way.

Keel — Keels are generally made from oak, though where a boat is to remain in the water most of the time, maple is much better. Mahogany, elm, beech and birch also make good keels. Keels are of various styles, according to the type of boat. Fig. 118 shows the construction of a small, shoal centerboard boat. Here there is no keel proper, the longitudinal strength being furnished by keelsons notched over the frames which are continuous from rail to rail, forward of and abaft the centerboard trunk. The port bedlog extends aft for a keelson, while both bedlogs are through fastened to the forward keelson, which is on the center line. Fig. 119 shows a somewhat similar construction for a small keel boat. Here all the frames are continuous and the keelsons extend far enough to get a good fastening into the deadwood. The strength amidships is furnished by the deadwood. The construction of larger keel boats is illustrated by Figs. 51 and 110. All the frames box into the keel, and transverse strength at the center line is preserved by floors at frequent intervals. The keel is deep enough to furnish the necessary longitudinal strength without a keelson.

Stringers, etc. — Stringers, shelf and clamp are usually of yellow pine. Oak is sometimes used and spruce where especial lightness is desired.

Deck — White pine is almost universally used for decks. Where it is finished bright, the seams are payed with some elastic seam composition after being caulked. Teak makes the finest deck possible for a yacht as it does not shrink or swell whether wet or dry. It takes a fine finish and may be oiled, varnished or holystoned. It also is splendid for skylights and deck trim in general. Yachts with a house generally have the deck planks sprung parallel with the planksheer. Where a yacht is flush decked, or has a house running straight fore and aft, like the 30-footer, the deck planks usually run straight, as this makes a little more shipshape appearance. Canvas-covered decks are used a great deal for small yachts, as a tight deck is thus secured with thinner planking than could be used otherwise. The deck planks should be matched boards of almost any light wood. A double diagonal deck with the two plies glued together makes an exceptionally strong deck.

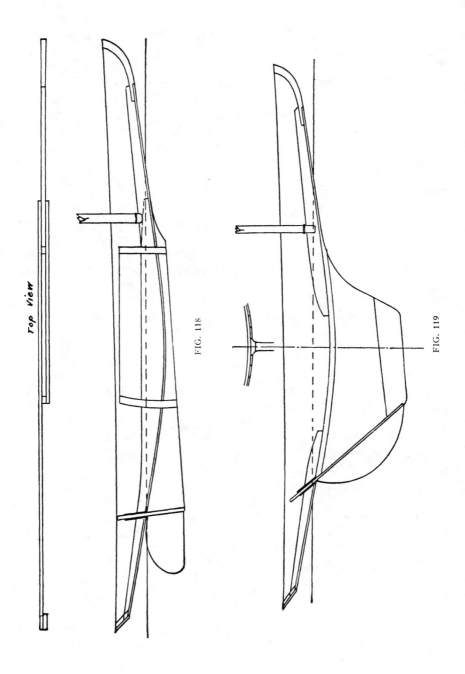

TOP VIEW

FIG. 118

FIG. 119

Deck Beams — Oak is commonly used for deck beams, though spruce may be used where extreme lightness is a desideratum. Beams are sawed to shape unless the crown is excessive, when they must be steamed and bent. Extra heavy beams should be located at masts, at ends of house and cockpit, and at skylights or other deck openings. These are termed main beams. Other intermediate beams continuous across the boat are called auxiliary beams, and the beams along each side of the house are known as half beams. A method of laying out a beam curve is shown in Fig. 120.

Miscellaneous — Floors are of oak, bronze or galvanized wrought iron. Bronze keel bolts should not land on iron floors, as galvanic action is liable to occur. Keel bolts are generally of Tobin or other strong bronzes which have an elastic limit of about 30,000 pounds. The lower end is commonly enlarged by forging to hold the lead, but it is better practice to thread and nut both ends, as forging reduces the

Beam Curve Construction

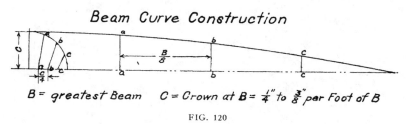

B = greatest Beam   C = Crown at B = $\frac{1}{4}''$ to $\frac{3}{8}''$ per Foot of B

FIG. 120

strength of bronze. Lloyd's Register gives complete rules for determining the sizes of keel bolts, based on width, depth and sectional area of the lead and on the spacing of the bolts. Customary working loads for various diameters of bolts are given roughly as follows:

Diameter $\frac{1}{2}''$, $\frac{5}{8}''$, $\frac{3}{4}''$, $\frac{7}{8}''$, $1''$, $1\frac{1}{8}''$, $1\frac{1}{4}''$, $1\frac{3}{8}''$, $1\frac{1}{2}''$. Load per bolt, lbs.: 200, 300, 440, 600, 790, 990, 1230, 1480, 1770. Lloyd's Table X allows for depth and width of lead and should be used.

Knees and stem are of natural crook oak or hackmatack. Tie rods and deck straps, as shown in Fig. 117, are most effective. The tie rods at the mast communicate a portion of the thrust of the mast to the main beams, thus relieving the keel. They also prevent the deck springing up under pull of the halliards. The deck straps are let into the beams and are well fastened to deck plank and beams. They resist the wringing tendency of the mast.

Longitudinal Construction — This construction is in extensive use on large steel vessels and has been used of late, to a slight extent, for small wooden yachts. It appears to offer some decided advantages. Briefly, it consists of web frames, or extra strong sawn frames, spaced far apart. On these are fitted longitudinal members on which the inner skin is laid diagonally and the outer skin fore and aft. The deck is double, with both skins diagonal. This makes a strong and rigid construction.

The foregoing covers very briefly the principal points of construction in wood. Table XI gives a representative schedule of scantlings for cruising boats of various water line lengths. The weights of various woods are given in Table II.

TABLE X

LLOYDS TABLE OF MINIMUM DIAMETERS OF BRONZE BOLTS FOR ATTACHING LEAD KEELS

Product of the sectional area of lead keel in square feet and the fore and aft spacing of keel bolts in feet	DIAMETERS						
	PROPORTION OF DEPTH OF LEAD KEEL TO TOP WIDTH						
	Under 1.0	1.0 and under 1.5	1.5 and under 2.0	2.0 and under 2.5	2.5 and under 3.0	3.0 and under 3.5	3.5 and under 4.0
	inches	inches	inches	inches	inches	inches	inches
Under .5	$\frac{9}{16}$	$\frac{9}{16}$	$\frac{9}{16}$	$\frac{5}{8}$	$\frac{3}{4}$	$\frac{7}{8}$	1
.5 and under .8	$\frac{9}{16}$	$\frac{9}{16}$	$\frac{5}{8}$	$\frac{3}{4}$	$\frac{7}{8}$	1	$1\frac{1}{8}$
.8 and under 1.2	$\frac{9}{16}$	$\frac{5}{8}$	$\frac{3}{4}$	$\frac{7}{8}$	1	$1\frac{1}{8}$	$1\frac{1}{4}$
1.2 and under 1.7	$\frac{5}{8}$	$\frac{3}{4}$	$\frac{7}{8}$	1	$1\frac{1}{8}$	$1\frac{1}{4}$	$1\frac{3}{8}$
1.7 and under 2.3	$\frac{3}{4}$	$\frac{7}{8}$	1	$1\frac{1}{8}$	$1\frac{1}{4}$	$1\frac{3}{8}$	$1\frac{1}{2}$
2.3 and under 3.0	$\frac{7}{8}$	1	$1\frac{1}{8}$	$1\frac{1}{4}$	$1\frac{3}{8}$	$1\frac{1}{2}$	$1\frac{5}{8}$
3.0 and under 3.8	1	$1\frac{1}{8}$	$1\frac{1}{4}$	$1\frac{3}{8}$	$1\frac{1}{2}$	$1\frac{5}{8}$	$1\frac{3}{4}$
3.8 and under 4.7	$1\frac{1}{8}$	$1\frac{1}{4}$	$1\frac{3}{8}$	$1\frac{1}{2}$	$1\frac{5}{8}$	$1\frac{3}{4}$	$1\frac{7}{8}$
4.7 and under 5.7	$1\frac{1}{4}$	$1\frac{3}{8}$	$1\frac{1}{2}$	$1\frac{5}{8}$	$1\frac{3}{4}$	$1\frac{7}{8}$	2
5.7 and under 6.8	$1\frac{3}{8}$	$1\frac{1}{2}$	$1\frac{5}{8}$	$1\frac{3}{4}$	$1\frac{7}{8}$	2	$2\frac{1}{8}$
6.8 and under 8.0	$1\frac{1}{2}$	$1\frac{5}{8}$	$1\frac{3}{4}$	$1\frac{7}{8}$	2	$2\frac{1}{8}$	$2\frac{1}{4}$

The scantlings of wooden racing yachts under 46 feet rating for Universal and International rules are regulated by Lloyds "R" rules or Herreshoff's "Rules for the Construction of Wooden Yachts" as published by the New York Yacht Club. These rules produce substantially constructed yachts and are applicable to cruising boats.

The composite construction is a combination of the all wood and all metal systems of construction and is used largely on yachts of from 50 to 90 feet water line. With this construction the frames, floors, reverse frames, keelson, stringers and deck beams are of metal, while the keel, planking and deck are of wood. Sometimes a part of the frames

TABLE XI

REPRESENTATIVE SCANTLINGS IN WOOD FOR KEEL BOATS

L.W.L.	Keel Sided	Stem Sided	Frames Spaced	Frames Sec. Area	Plank Thickness	Deck Thickness	Deck Beams Sec. Area	Shelf & Clamp Sec. Area	Stringers Sec. Area
15'	7"	2¾"	8"	.9"	5/8"	3/4"	1"	3"	4"
18	8	3	9	1.0	3/4	7/8	1.3	4	4
21	9	3¼	10	2.3	7/8	1	1.7	8	6
25	11	3¾	12	3.3	1	1⅛	2.3	12	8
30	13	4¼	12	4.8	1⅛	1¼	3.0	16	10
35	14	5	12	6.0	1¼	1⅜	3.7	20	12
40	15	5½	14	7.7	1⅜	1½	4.4	24	14
45	16	6	16	11.	1½	1⅝	5.1	28	16
50	17	6½	18	14.	1⅝	1¾	6.5	31	18
60	19	7½	22	21.	1⅞	2	13.	37	23
70	21	8	24	28.	2⅛	2¼	19.	43	27
80	23	9	24	36.	2⅜	2⅜	22.	49	31
90	25	9½	24	43.	2⅝	2½	24.	55	35

Deck beams are spaced the same as the frames.

and deck beams are of wood also. Fig. 121 is the midship section of a 75-foot schooner and is typical of the composite construction. A flanged steel plate is bolted on top of the wood keel and to this the

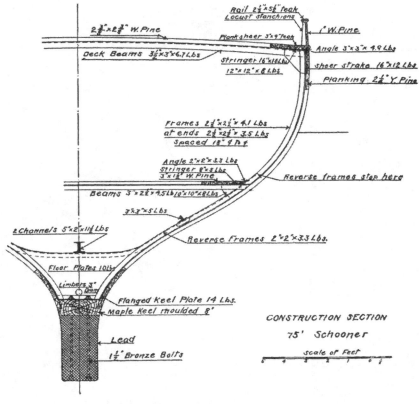

FIG. 121

frames are riveted. Floor plates, clips and reverse frames complete the strength at the center line.

The wood keel is of maple, eight inches thick. The lead is cast in one piece and the bolts pass through the keel and set up on the inside of the keel plate. Other features are similar to those of the all metal construction.

CHAPTER XV

RESISTANCE

A KNOWLEDGE of resistance is of fundamental importance in the design of mechanically propelled craft. On the definite knowledge of the amount of resistance depends the determination of the power necessary to produce a desired speed and the proportions of the propeller, or other means of applying the power. This is a large subject and I can do no more in this book than outline the principal sources of resistance and the methods of determining their amount.

There is not much point in determining resistance in the case of a yacht with sail as the motive power; the wind is indeterminate in force and can be made use of in limited degree only. With the large displacement necessary for sail-carrying power, the possible speed range is also extremely limited as compared with mechanical propulsion. 1.4 is about the maximum speed-length ratio possible with seagoing sailing yachts, which is less than a quarter of the tentative limit of power-driven yachts of non-planing type.

The resistance met by a vessel in motion is of several kinds and includes frictional resistance, wave-making resistance, eddy-making resistance and air resistance.

SKIN RESISTANCE

Skin or frictional resistance is the largest source of resistance for low or moderate speeds. It is due to the fact that water is not frictionless, in consequence of which every square foot of submerged area experiences a certain amount of drag in passing through the water. Fig. 122 shows a series of curves indicating the fraction of the total resistance due to the skin friction for a 50-foot power boat design expanded to various displacement-length ratios. It will be noted that the smaller the displacement-length ratio, the higher the percentage of total resistance due to skin friction; in other words, the wave-making resistance is proportionately less in the finer models at all speeds.

The celebrated English physicist, William Froude, investigated this subject thoroughly many years ago and present-day methods of estimating skin resistance are based on his work.

The formula for frictional resistance of plane surfaces passing through the water is

$$R_f = fSV^n,$$

where f is the coefficient of friction of the surface, S is the total area of wetted surface in square feet, V is the speed in knots, and n is an exponent, the value of which for general work is 1.83. This exponent actually varies considerably with length and smoothness of the surface, but 1.83 is a good all round value for long smooth surfaces. The value of f, the coefficient of friction, varies with the length of the surface due to the fact that the forward part of the surface enters undisturbed water whereas the after portion passes through water that has had some motion imparted to it; the longer the surface, the greater the

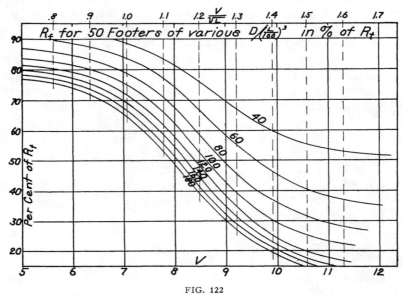

FIG. 122

fraction of the total area which is in contact with water which has acquired forward motion.

Fig. 123 gives Froude's values of f for smooth painted surfaces of a wide range of lengths.

Where the bottom is very foul, this coefficient f may easily be doubled. Knowing the area of under water surface of the yacht, and her speed, we can readily calculate the resistance due to skin friction by the above formula.

In addition to skin resistance of bare hull, we have to consider the resistance of rudder, shaft struts, shaft, external or bilge keels, etc., all of which have a resistance per square foot greater than that of the hull proper. For the rudder, a suitable correction is to double the actual area to allow for eddy making, because of the greater speed of the

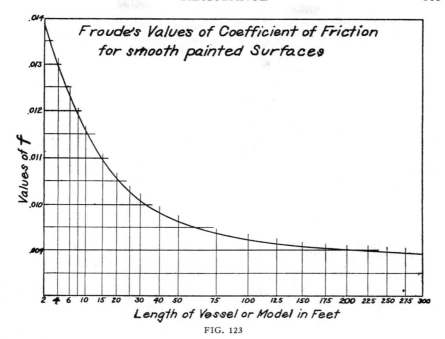

FIG. 123

TABLE XII

VALUES OF $V^{2.83}$ AND $V^{1.83}$

V	$V^{2.83}$	$V^{1.83}$	V	$V^{2.83}$	$V^{1.83}$
3	22	7.4	13	1421	110
3.5	35	9.9	13.5	1581	118
4	51	12.6	14	1752	125
4.5	70	15.7	14.5	1935	133
5	95	19.0	15	2130	142
5.5	124	22.8	15.5	2337	151
6	159	26.6	16	2557	160
6.5	200	30.8	16.5	2789	169
7	246	35.2	17	3055	178
7.5	300	40.0	17.5	3295	188
8	360	44.9	18	3868	199
8.5	427	50.1	18.5	3856	209
9	502	55.6	19	4158	219
9.5	585	61.5	19.5	4475	229
10	676	67.6	20	4808	240
10.5	776	73.9	20.5	5156	251
11	885	80.7	21	5519	263
11.5	1004	87.3	21.5	5899	274
12	1113	94.4	22	6296	285
12.5	1271	102	22.5	6709	297

water at this point due to the propeller race and the usual slight helm angularity. Struts well placed so as to lie fair in the stream lines may have their actual surface area multiplied by four to give a good approximation of their resistance.

WAVE-MAKING RESISTANCE

At considerable speeds the resistance due to the formation of waves is important, and at high speeds quickly runs up to amounts which render further increase in speed extremely costly. Wave-making resistance is best measured in pounds per ton of displacement. We have available a large amount of data on this quantity which have been ascertained by extensive experiments with models tested by Froude, Admiral D. W. Taylor and others for a wide variety of displacement-length ratios, speed-length ratios, prismatic coefficients and various ratios of proportions. Results of experiments with Taylor's standard series of 80 models constructed from one parent design is summarized in Fig. 124. This gives resistance per ton for displacement-length ratios of 40, 60, 80, 100, 120, 140, 160 and 180, all based on the model having the most suitable prismatic coefficient for the given speed. In the same figure, the value of the best prismatic coefficients to use is indicated in the dotted curve. It is most important, by the way, for best efficiency to use the proper prismatic coefficient. For instance: the wave-making resistance for

$$\frac{V}{\sqrt{L}} = 1 \text{ and } \frac{D}{\left(\dfrac{L}{100}\right)^3} = 100$$

is 4 lbs. per ton with best prismatic coefficient = .55. For a prismatic coefficient of .65, this resistance is doubled. It will be noted that the higher the speed, the greater the prismatic coefficient, which means that the displacement is worked more towards the end of the vessel since the middle body is a great wave-making factor at high speed. The curves in Fig. 124 are for a moderate speed model of particularly favorable form and, of course, do not apply actually to any other model. The results judiciously used are, however, sufficiently close for estimating resistance of normal hulls at medium speeds to give a close approximation of the wave-making resistance. It must be borne in mind also that wave-making resistance is only a fraction of the total resistance — a large one, it is true, for speed-length ratios of over 1.2 in heavy models — so that an error in R_w means a smaller percentage error in R_t. Taylor's values of R_w will not, as a rule, be found to underestimate the resistance. In finding the total wave-making resistance of the vessel for the particular speed and displacement-length ratio, we have simply to multiply the resistance per ton at that speed by the displacement in tons.

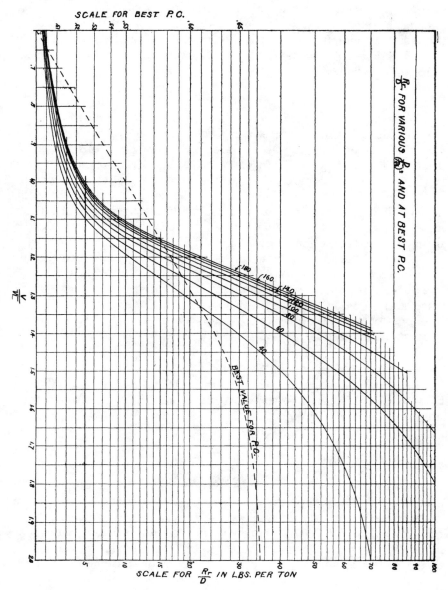

FIG. 124

EDDY-MAKING RESISTANCE

Eddy-making occurs at propeller apertures, at the after edge of thick rudders and at any other subsurface appendages or projections that are not neatly streamlined. There is little excuse for much resistance of this kind in a well designed yacht and it is commonly ignored, the factor by which the effective horse power is multiplied to get the indicated horse power being taken large enough to allow for such source of resistance.

AIR RESISTANCE

The passage of a yacht through the air sets up resistance in the same way that it does through the water, but as air is only about 1/800 the weight of water, the resistance under ordinary conditions is a small fraction of the water resistance. When the relative speed of a boat through the air is 30 knots or more, this resistance becomes important. Froude investigated air resistance as well as water resistance in his famous *Greyhound* experiments and found that air resistance, without masts or rigging, at 10 knots was about 1½ per cent of the water resistance. A formula for approximating air resistance for a vessel is $R_a = .0043\ AV^2$, where A is the cross sectional area above water and V speed in knots.

The resistance per ton given in Fig. 124 is really residuary resistance, since eddy-making and air resistance are included with the wave-making resistance. However, the terms wave-making and residuary resistance are often used interchangeably.

HORSE POWER

The actual horse power used in propulsion, called the "effective horse power," is equal to the total resistance times speed in feet per minute divided by 33,000 or:

$$E_t = \frac{R_t \times 6080\ V}{60 \times 33000} = .00307\ R_t\ V \qquad R_t = R_f + R_w.$$

Since there are various serious losses between the engine and the actual useful power applied, it is necessary to divide the effective horse power by the coefficient of propulsion to get the indicated horse power. This coefficient is approximately .5 in ordinary cases and allows for propeller losses, friction in bearings, power consumed by direct driven auxiliaries and internal friction in the engine. For the usual motor or auxiliary cruiser, the effective horse power is about half the brake horse power.

40' "TEASER" Crouch

For quick approximation of power, I have prepared a series of curves shown in Figs. 126 to 144. These are based on the calculation of power required for a wide range of modifications of the parent design shown in Fig. 125, including eighteen different lengths and eight displacements on each length obtained by expanding beam and depth proportionately the amount necessary to give the desired displacement-length ratios. Only one set of lines was drawn, but after finding the displacement and wetted surface for this design the wetted surface for lateral expansions or contractions was calculated, using the principle that area varies as $\sqrt{D}$ under these conditions. For larger and smaller boats having the same displacement-length ratios, the area S (hull

26'   "DOLPHIN"   Hacker

only) varies as L^2, and displacement as L^3. The areas of rudder and keel on a given L were varied only slightly to be suitable for the given displacement.

This series of curves, which involved a large amount of calculation, is useful for first approximations of power and for studying the effect of varying the power or weights in a given hull. In preparing a preliminary design, the client often wishes to know what speed a certain engine will give. Also, he may want to know how much increase in speed he will get by fitting another engine of greater power or what speed he would get if he lengthened the hull a certain amount. In the case of increased power, the new displacement-length ratio will be calculated if the displacement is increased and the power read from the curve for the speed corresponding to this displacement-length ratio. If the hull is stretched out merely by spacing the stations farther apart, the displacement will be increased in direct proportion to the length.

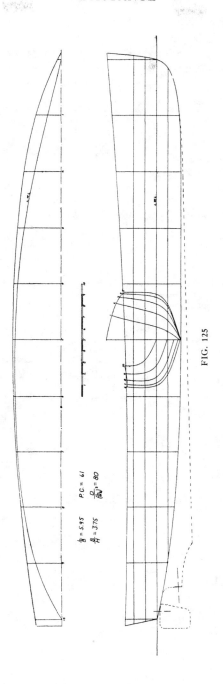

FIG. 125

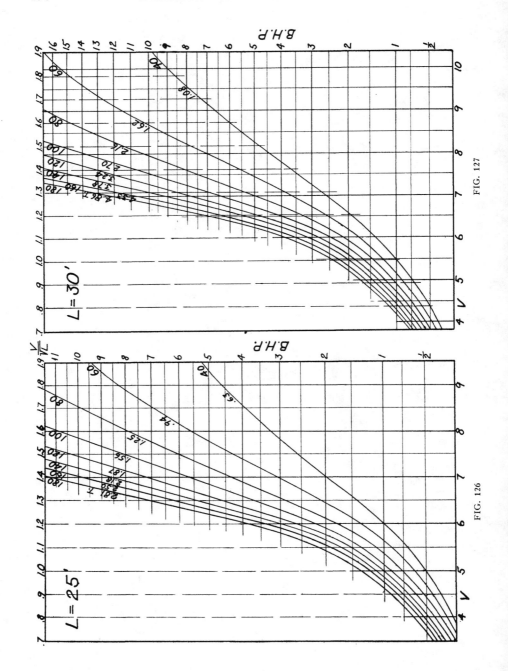

FIG. 127

FIG. 126

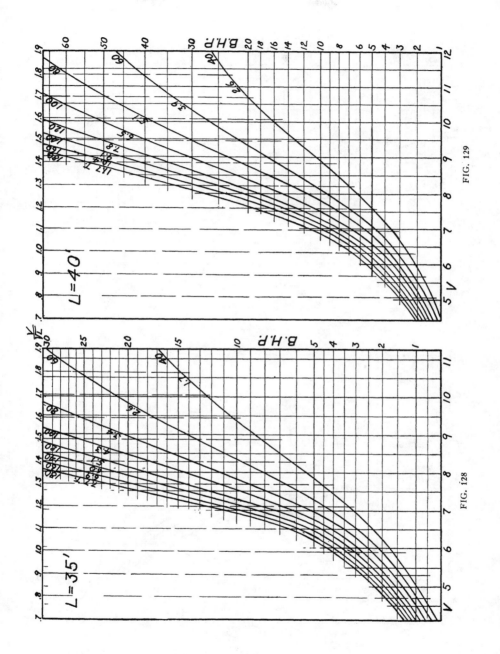

FIG. 129

FIG. 128

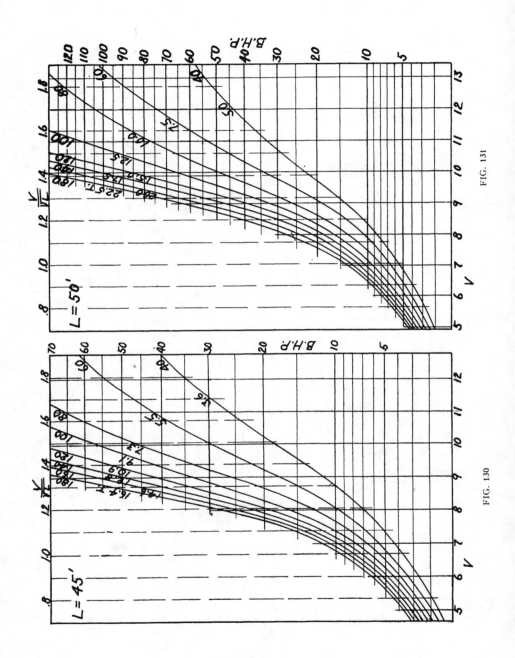

FIG. 131

FIG. 130

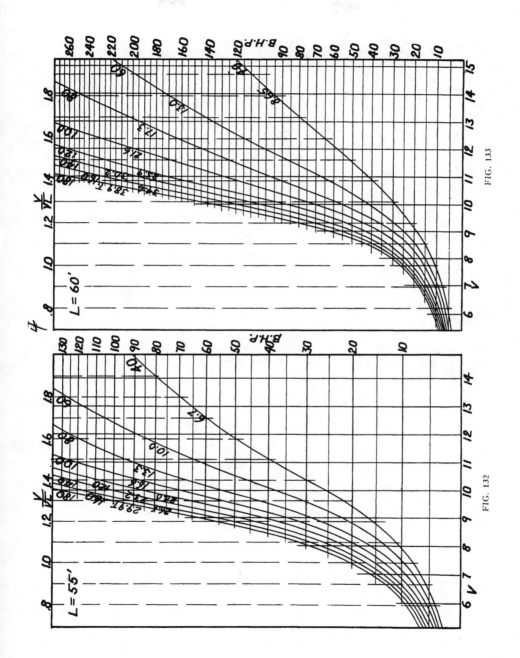

FIG. 133

FIG. 132

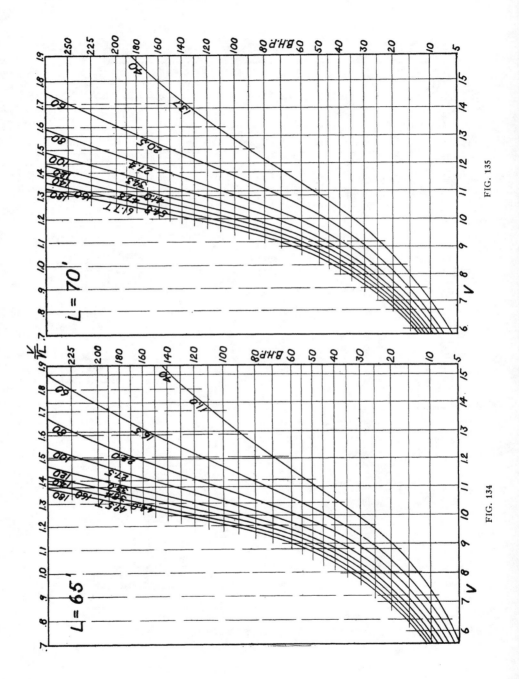

FIG. 135

FIG. 134

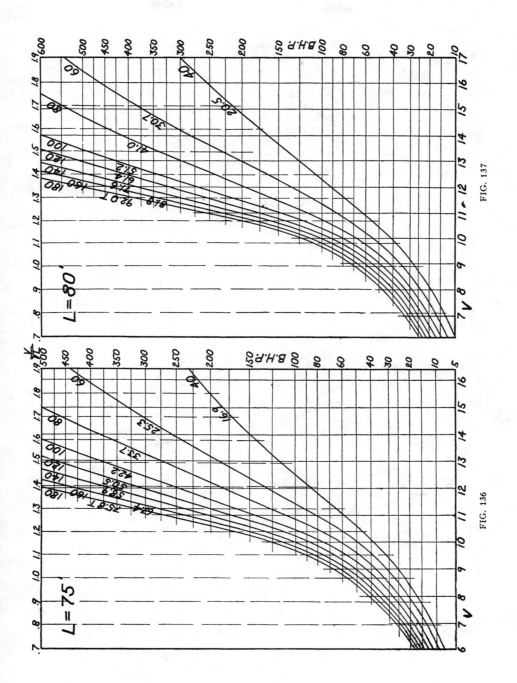

FIG. 137

FIG. 136

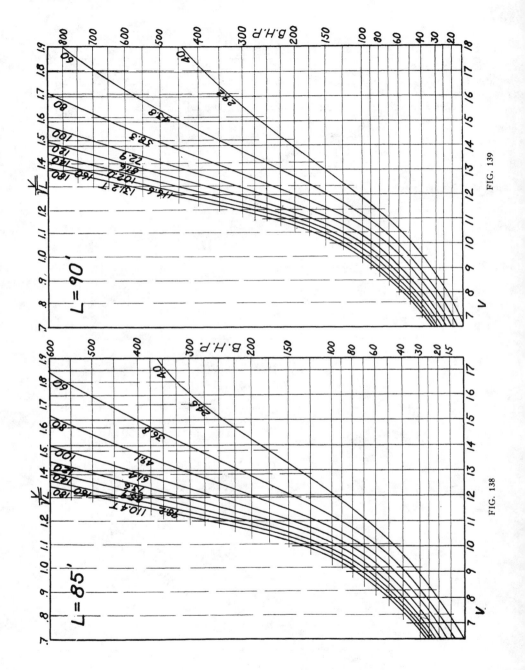

FIG. 139

FIG. 138

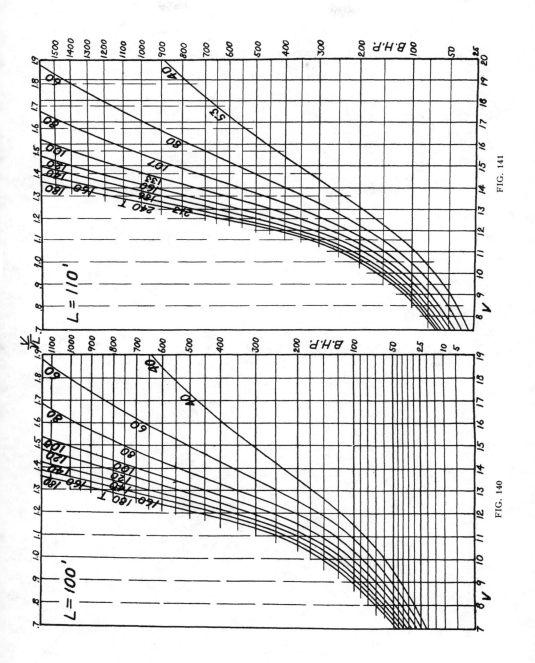

FIG. 141

FIG. 140

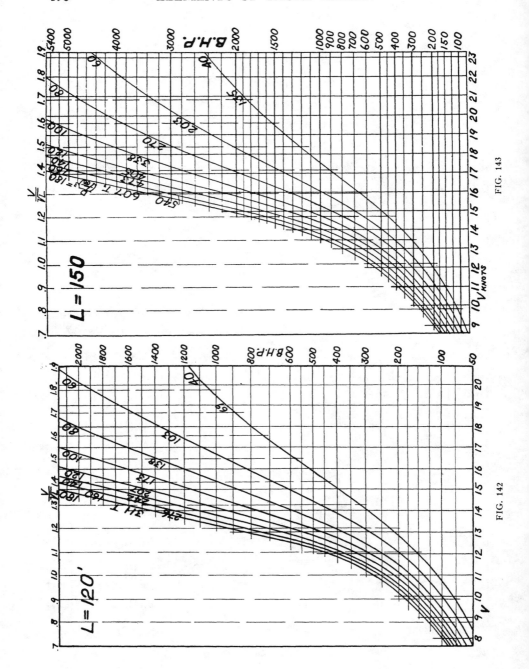

FIG. 143

FIG. 142

The curves will be found useful for many purposes. The displacement-length ratio and the displacement in tons of 2240 lbs. are indicated on each curve.

Where the displacement-length ratio, as is generally the case with

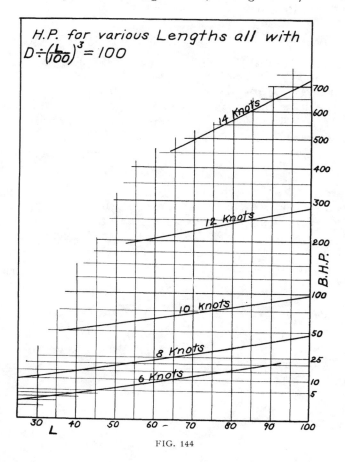

FIG. 144

auxiliaries, is greater than 180, the horse power for 180 displacement-length ratio is to be multiplied by the ratio of the displacement-length ratio to 180. Take the case of a 60-foot auxiliary having a displacement-length ratio of 275, for which a speed of 9 knots is desired. The horse power for 180 displacement-length ratio is 65; multiplying this by 275/180 the required horse power is 100.

In preparing these curves the calculated effective horse powers were

multiplied by 2 to get the brake horse power. In the case of a very efficient installation this will give too great horse power; in some instances, such as auxiliaries with thick deadwoods and rudders, it will
underestimate the power. Special circumstances must therefore be
taken into consideration.

DIRECT CALCULATION OF HORSE POWER

According to the theory of similitude, the wave-making resistance for
similar hulls at corresponding speeds is directly proportionate to the
displacements; knowing the resistance per ton for a given model, a
vessel of the same model, no matter what the size, will have a total
wave-making resistance equal to this resistance per ton multiplied by
the number of tons displacement, so that

$$R_w = \frac{R_w}{D} \times D$$

In making power calculations it is desirable to figure the power
necessary to drive the vessel at various speeds below the desired maximum and to draw curves of power necessary to drive vessels at any
given speed. Such calculation is illustrated below.

HORSE-POWER CALCULATION

BASED ON TAYLOR'S STANDARD SERIES DATA

$$L = 75 \qquad D = 33.7\frac{D}{\left(\frac{L}{100}\right)^3} = 80 \qquad S = 950$$

V	$V^{2.83}$	P_f	$\frac{V}{L}$	$\frac{R_w}{D}$	P_w	P_t
6	159	8.7	.69	1.1	1.4	10.1
7.5	300	16.5	.76	2.1	3.3	19.8
9	502	27.5	1.04	4.5	8.4	35.9
10.5	776	42.6	1.21	17	37.0	79.6
12	1133	62.1	1.39	48	119.2	181.3
13.5	1581	86.6	1.56	83	232.0	318.6

$$P_f = f \times S \times V^{2.83} \times .00307 \div \text{eff. of propulsion}$$
$$= .0094 \times 950 \times V^{2.83} \times .00307 \div .5 = .0548 \ V^{2.83}$$

$$P_w = \frac{R_w}{D} \times D \times V \times .00307 \div \text{eff. of propulsion}$$

$$= \frac{R_w}{D} \times 33.7 V \times .00307 \div .5 = .207 \frac{R_w}{D} \times V$$

$$P_t = P_f + P_w$$

The case chosen for illustration is that of a 75-footer having a displacement-length ratio of 80. The speeds for which the horse powers are to be calculated are recorded in the first column, and the 2.83 power (see table XII) of these speeds is noted in the second column. Horse powers absorbed by skin friction are calculated independently and recorded in the third column. After figuring the speed-length ratios for the various speeds, the corresponding wave-making resistances per ton are taken from Fig. 124 and recorded in the fifth column. The wave-making horse powers are then calculated. These, added to the frictional horse powers, give total horse powers required as recorded in the last column. A curve may then be plotted which will give horse powers for all speeds from 6 to 13.5.

POWER CALCULATION BASED ON MODEL EXPERIMENTS

Where a design represents a departure from existing designs on which complete data is available, or when extreme accuracy is desired, the proper method of determining resistance is by the use of a scale model constructed on the exact lines of the vessel. This model is towed in a tank and the resistance measured at a range of speeds up to a maximum corresponding to the maximum desired speed of the vessel. The theory of mechanical similitude applies only to residuary resistance since the area of wetted surface does not vary as the cube of the length and since the coefficient of friction also varies more or less in proportion to the length. It is necessary, therefore, to separate the resistance of the model into its components of frictional and wave-making (or residuary) resistance, which latter is obtained by subtracting the calculated skin resistance from the total resistance. This process is best illustrated by the calculation given below.

HORSE-POWER CALCULATION

BASED ON MODEL TEST

	L	D	$\dfrac{D}{\left(\dfrac{L}{100}\right)^3}$	S
Model	5	.0125	100	4
Yacht	50	12.5	100	400

V_m	R_t	$V_m^{1.83}$	R_f	R_w	V_s	P_w	$V_s^{2.83}$	P_f	P_t	$\dfrac{V}{\sqrt{L}}$
3	1.08	7.4	.38	.70	9.48	41.8	582	13.9	55.7	1.34
3.5	2.19	9.9	.50	1.69	11.07	117.5	901	21.4	138.9	1.56
4	3.20	12.6	.64	2.56	12.65	203.7	1315	31.3	235.0	1.79
4.5	4.20	15.7	.80	3.40	14.23	304.2	1835	43.7	347.9	2.01

V_m = Speed of model in knots

R_t = Total resistance of model in lbs.

R_f = Frictional resistance of model in lbs.

$\quad = f \times S \times V_m^{1.83} = .0127 \times 4 \times V_m^{1.83} = .0508 V_m^{1.83}$

R_w = Wave making resistance of model $= R_t - R_f$

V_s = Speed of yacht in knots $= V_m \times \sqrt{\dfrac{50}{5}} = 3.16 V_m$

P_w = Brake hp. of yacht to overcome wave making

$\quad = R_w \times \dfrac{64}{62.4} \times \dfrac{D_s}{D_m} \times .00307 \times V_s \div$ eff. of propulsion $=$

$\quad \dfrac{1.025 \times 1000}{.5} \times .00307\, R_w V_s = 6.29\, R_w V_s.$

P_f = Brake hp. of yacht to overcome skin friction

$\quad = f \times S \times V_s^{2.83} \times .00307 \div$ eff. of propulsion

$\quad = .0097 \times 400 \times .00307 \div 5 \times V_s^{2.83} = .0238 V_s^{2.83}$

$P_t = P_w + P_f$

Here the horse powers to drive a 50-foot hull at speeds of 10 to 14 knots are to be derived from resistance tests on a 1/10 size model. The speed range necessary for the model is that of the yacht divided by $\sqrt{10}$. The column R_t gives the total resistance of the model by test at the various speeds. The frictional resistances are then calculated and subtracted from the total resistances, which leaves residuary resistances R_w. The speeds of the yacht V_s, corresponding to the model speeds, are the latter multiplied by $\sqrt{10}$. The residuary resistances of the yacht are those of the model $\times$ 10^3 which, multiplied by

$$.00307 \times \frac{64}{62.4} \times V_s$$

divided by efficiency of propulsion which is assumed to .5, give residual horse powers, P_w. If the model is tested in fresh water and the yacht is to be tested in salt, it is necessary to multiply by the ratio of the densities of salt and fresh water, which is the second factor in the equation above. P_f, the skin friction horse power, is calculated as already explained. The ratio of surface areas of yacht and model is

$$\left(\frac{L}{1}\right)^2 \text{ or } \left(\frac{D}{d}\right)^{2/3} = 100.$$

Carefully conducted model tests give the only accurate basis for predicting power requirements of new models and are decidedly much worth while in working out the design for an expensive yacht. The designer can get these tests made for him in the Government basin at Washington, or in the University of Michigan's basin at Ann Arbor; at Stevens Institute in Hoboken, N. J.; at New York University; or can run them himself in suitable sheltered water. This matter is a subject that should be of great interest to the progressive designer.

PROPULSIVE COEFFICIENT

After determining the different horse powers required to produce a given speed, the corresponding brake horse power is obtained by dividing $E H P$ by the propulsive coefficient. The latter is the product of the propeller efficiency, the efficiency of intermediate shaft bearings and a factor allowing for the obliquity of the stream lines with reference to the axis of propeller. These factors are roughly, under average conditions, .60, .90, and .90, which gives a total efficiency slightly less than .50 which means that the brake horse power of the engine must be at least twice the effective horse power.

HORSE POWER BY COMPARISON

Where data is available on a vessel of quite similar form and not widely dissimilar size, the speed of the new design may be approximated by comparison with the old. The horse power of the new will be equal to the horse power of the old multiplied by the ratio of the displacements to the 7/6 power. This is, of course, at the corresponding speed and assumes that the total resistance varies as the displacements, which is not strictly accurate as explained previously. For a speed different from the corresponding one, it will be necessary to multiply by the ratio of the new speed to the old, raised to a certain power. An idea of the proper exponent to use may be gotten from Fig. 145, which shows the power of the speed at which the horse power varies for various displacement-length ratios for the design shown in Fig. 125. Figs. 21 and 22 give values of $D^{7/6}$.

An interesting figure is 143, which shows power necessary to obtain various speeds for different length boats all of displacement-length ratios of 100.

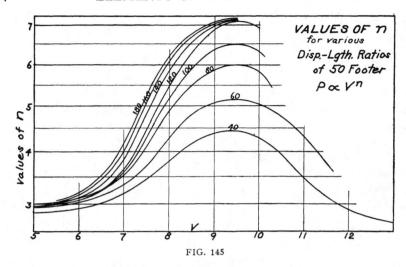

FIG. 145

Another interesting diagram is Fig. 146, which shows the ratio of horse power to the 7/6 power of the displacement for various displacement-length ratios and speed-length ratios. This ratio is a fair index of hull efficiency and the curves show that the finer models have least efficiency by this standard at speed-length ratios up to 1.4. Above this speed-length ratio the bulkier models become less efficient due to the great increase in wave-making resistance. These curves were calculated for 50-foot length, and greater lengths show slightly

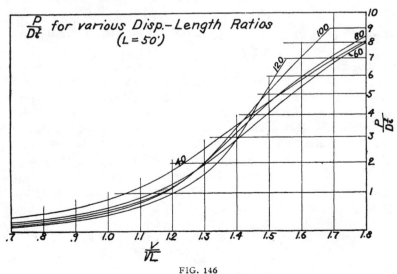

FIG. 146

better efficiency due to the fact that wetted surface for similar hulls increases only as the square of the length whereas the wave resistance increases as the displacement or as the cube of the length. However, the fact that most of the curves lie close together indicates a short cut in figuring powers up to the limits given in Figs. 147 and 148. To illus-

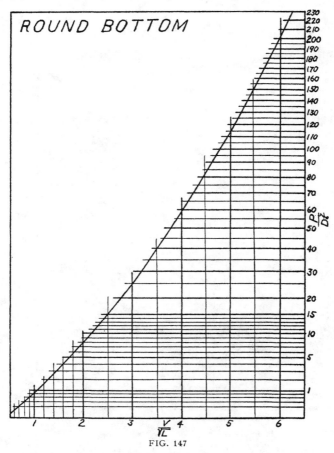

FIG. 147

trate the method of using these curves, what will be the approximate horse power required to drive a round bottom boat of 80′ *L.W.L.* and 40 tons displacement at a speed of 20 knots?

$$D^{7/6} = 70 \text{ (Fig. 22)} \quad \frac{V}{\sqrt{L}} = 2.24$$

From Fig. 147 $\frac{P}{D^{7/6}} = 12$ and $P = 12 \times 70 = 840$

SHAPE OF MIDSHIP SECTION

A semicircular section involves least amount of wetted surface and the least deflection of stream lines. Considerations of stability, seaworthiness, internal accommodations and draft, however, usually demand a different section. Small power craft have a much larger ratio

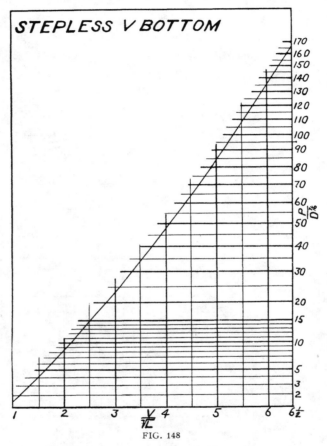

FIG. 148

of beam to length than larger vessels, as narrow beam entails correspondingly less height of side, and both beam and depth are necessary to get the requisite accommodations. A round bilge with a moderate deadrise and straight bottom near the center line are the best combination for the small yacht. Larger vessels have flat bottoms and hard bilges to get the necessary displacement on the given beam and draft, and such a section is not inefficient when associated with a large ratio of length to beam.

SHAPE OF THE LOAD WATER LINE

For speed-length ratios up to 1.5, a fine water line forward, with some hollow, is desirable. Above this speed-length ratio the hollow is best eliminated. Round sections — that is, fairly full ones rather than V sections — are best forward, always being shaped to give the proper areas as indicated by the ideal area curve constructed in advance. The shape of the after sections is not so important. In general, the after portion of the water line should be full and the depth of sections less

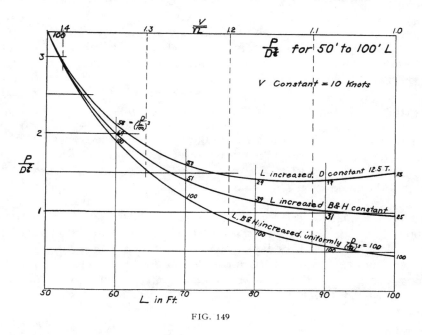

FIG. 149

than forward or well fined in a reverse curve if carried down to a deadwood.

ADVANTAGE OF LENGTH

In the moderate speed types such as are discussed in this chapter, length is the most important function for speed. The low speed-length ratio has the least resistance per ton for wave-making. Fig. 149 shows in an interesting way the effect of adding length to a 50-foot cruiser having a displacement-length ratio of 100 and a speed of 10 knots.

The three curves show the value of the ratio $\dfrac{P}{D^{7/6}}$.

The lower curve is for the yacht increased in size proportionately in all three dimensions, so that the displacement-length ratio is 100 throughout. As the length increases, the displacement is increased as the cube of the length and the wetted surface as the square. The frictional resistance is increasing directly as the area of wetted surface but the resistance per ton for wave-making is constantly and greatly decreasing as the speed-length ratio decreases, so that the optimum length is far beyond 100 feet. The middle curve shows the length increased with the beam and draft constant, that is, the boat is stretched out by spacing the stations farther apart. The saving in wave-making resistance on account of the increasingly lower speed-length ratio is not offset by the increasing frictional resistance within the limits of this curve. The figures at each 10 foot length indicate the displacement-length ratios for that length. The upper curve is most interesting of all and shows how the conflict between decreasing wave-making resistance and increasing frictional resistance comes out when the displacement is kept constant and beam and draft are reduced as the length increases. The unit wave-making resistance decreases so slowly that the frictional resistance overcomes the saving in wave-making resistance at a speed-length ratio of about 1.1, which indicates that 88 feet is the maximum economical length for the given displacement of 12½ tons. At this length the displacement-length ratio is about 20, which is much lower than is desirable for other considerations and much lower than is ever used in practice, with the exception of racing shells. In the case of the latter, the displacement-length ratio is much lower than this, but there are other considerations demanding great length beside that of least resistance. Length, while desirable for speed, is undesirable from almost every other standpoint and a compromise must be made on this point.

In naval architectural calculations, speed is always expressed in knots instead of statute miles. This being the universal practice, it is preferable to use knots always in expressing speed. Motor boat owners generally use statute miles per hour as a speed unit largely because small motor boats are extensively used inland, far from the sea and its customs, but also, we suspect, because speed in this unit looks considerably larger than in knots.

A matter of great importance in connection with the speed of yachts is the depth of water. In shallow water, the normal wave formation is affected, with consequent influence on the wave-making resistance. In extremely shallow water, the resistance may be considerably decreased at certain speed-length ratios, thereby enabling the yacht to make actually better speed than in deep water. At other depths the resistance may be considerably increased. In making speed trials, it is important to have adequate depth of water to avoid the possibility of an abnormal speed being obtained. The least depth of water in feet for avoiding a false speed is indicated by the formula

$$10H - \frac{V}{\sqrt{L}},$$

H being the submerged depth of hull in feet.

CHAPTER XVI

THE HYDROPLANE

FIFTY years before engines of high power and light weight were available, the basic idea of a hull with a bottom formed to skim over the surface of the water instead of going through it was invented and reinvented a dozen times. Early attempts to build a hydroplane or skimming hull were unsuccessful because engines were so heavy that no boat could float the power required to drive her at a speed high enough to plane. The idea is old but its practical application began about 1910.

It has been shown in previous chapters how an ever increasing percentage of the total resistance to a boat's motion is caused by the wave-making of the hull of normal form. It has been shown how the resistance increases so rapidly at high speed for a small increase in speed that speeds of boats of the displacement type are definitely limited and depend on the lengths. This does not hold true of the hydroplane where the bottom is so formed that forward motion at high speed causes the whole boat to rise bodily and to be supported by the dynamic reaction between the water and the bottom of the boat.

All ideas of stability as figured for normal forms are null and void. All major forces are entirely dynamic in character and the hydroplane designer must think of them as such. On the bottom of the boat, all surfaces and rounded shapes which might cause suctions must be avoided. In other words the designer must think of impacts, reactions and of water leaving edges cleanly and sharply rather than think of flow around a body in the water. When the resistance of a hull which is completely planing on the surface of the water is analyzed, all the various forms of resistance previously mentioned are found and in addition there is another and important factor.

The hydroplane has frictional resistance on her actual wetted surface. The displaced water creates wave-making resistance and there is eddy resistance behind the shafts, struts, rudders and other appendages. In addition, there is a horizontal resistance which is the horizontal component of the perpendicular pressure on the inclined planing surface of the bottom.

Fig. 150 represents a simple hydroplane moving along the surface of the water in the direction shown by the arrow. No matter how high the boat planes, there must be water displaced with consequent wave-making resistance and, although the area is greatly reduced compared to the surface in contact with the water when the boat is at rest, yet there is an appreciable amount of wetted surface with its frictional resistance. Rudders, struts, skid fins, propeller hubs, water scoops, bailers and other hull projections, each adds to the total resistance.

In the diagram, the water pressure is shown acting upward from A to P perpendicular to the bottom. The point P is the center of pressure and must be the point about which the boat balances when lifted. If this center is too far aft, the bow will drop; if too far forward, the bow will rise. One of the commonest faults of a hydroplane is a "porpoising" or leaping motion caused by the weight of the boat, the thrust of

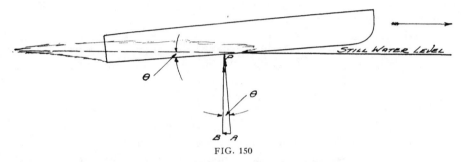

FIG. 150

the screw and the lift of the planes being improperly related one to another. Unfortunately, the actual location of the point of lift cannot be determined by a simple inspection of the lines of the hydroplane, nor can it be given by any mathematical formula known today. This normal upward thrust of the plane is the resultant of many elementary pressures and suctions. Even a true plane moving at an angle to the water has strong downward suctions instead of lifts at the after end. In many cases, the actual C.G. of the whole boat may be carried ahead of the point where the keel touches the water, thus creating the curious paradox that the center of the resultant upward pressure on the bottom of the boat is ahead of the point of contact of the bottom with the water surface. An upward curve at the after end of the bottom tends to exaggerate this condition. Bottom lines, therefore, should avoid any suspicion of upward curvature at the stern if the hull is to be a successful hydroplane.

The normal upward pressure AP can be resolved into two components, AB which is horizontal and opposed to the forward motion of the boat thus acting as an added resistance, and BP, a vertical component equal to the total weight of the boat.

From the diagram, it is evident that $AB = BP \times$ tangent θ. For a given value of BP, equal to the weight of the boat, AB decreases as the angle decreases. At first sight, it might appear that to make the added resistance very small, by keeping the angle θ small, would produce the most efficient hydroplane. This, however, is not the truth and, although both this horizontal component and the wave making resistance decrease as the angle θ decreases, the frictional resistance increases due to the added amount of surface required for lifting the boat at the small angle. By mathematical treatment it can be shown for a plane surface that the angle θ should be approximately $2\frac{1}{2}°$ to create the least possible resistance, which agrees well with angles used on successful hydroplanes. It is advisable to make the angle of the plane less rather than greater if any departure from the conventional angle is contemplated.

Little accurate information can be obtained from model tank experiments on small model hydroplanes as to the actual resistance, speed or the action at speed of a full sized boat. Model tests can and do give comparative results only, and then only if carefully interpreted. To attempt to divide the total resistance of a hydroplane into its elements as is done with ships and their models is, in the present state of our knowledge, not at all practical and is definitely misleading. As far as practical results are concerned, it can be assumed that the resistances of the model and boat have the same ratio as the displacements of the two. Any attempt to separate the resistance of the model into frictional resistance and the wave-making resistance, by subtracting the calculated frictional resistance of the model from the total resistance, requires an accurate estimate of the area of the model in contact with the water. The error in making such estimates on small models and in determining the proper coefficients to use for the frictional resistance is far greater than the error involved in assuming that resistances vary directly as the displacements at corresponding speeds.

Under average conditions, the surface tension of water is a constant. This constant determines the size of drops which water will form and this size is exactly the same, both for the model and for the boat. Surface tension will hold the small bow wave of a model together in a thin glassy sheet. On the boat, this wave will break up into drops or white spray. On the model, the glassy bow wave will adhere to the boat and may cause a suction which entirely changes the trim, the amount of wetted surface and the resistance of the model from that found in the boat.

The propeller thrust, too, being a fairly large percentage of the total weight, is another force which causes the model and boat to act differently; the change of flow in the water streams beneath the bottom, caused by the propeller accelerating water to it, also upsets the mathematical relation between a motor driven boat and a towed model. Many naval architects of wide experience on displacement boats have been misled by misinterpretation of model tank results on hydroplanes. The safest procedure to follow in developing a hydroplane design is to make a model of a known and successful boat, one whose performance, characteristics, speed, power, and weight have been accurately determined. This model can be used as a check on a model embodying any proposed design or improvement. But even

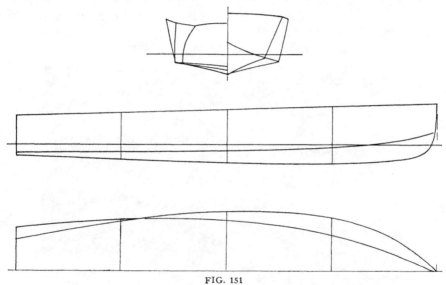

FIG. 151

this must be used with caution for cases have occurred where the model of a successful boat, a record breaking champion, could not be towed in the tank at speeds corresponding to the official speeds which the boat had actually made.

Models eight feet, ten feet, or twelve feet in length would give practical results if they could be towed in a high speed tank. Although such a tank is owned and operated by the United States Government, it is not available to the private naval architect and, until the services of this tank can be obtained, hydroplane design will continue to be a hit and miss rule of thumb affair where success depends more on small and careful experimental changes in the bottom of the actual boat and in the perfect tuning up of the motor and on propeller experiments than on any truly scientific design based on exact knowledge.

Modern hydroplanes are of six general types. They are as follows:

1. Stepless hydroplanes.
2. Multiple-step or "shingled" hydroplanes.
3. Single step.
4. Inverted "V" hydroplanes or "Sea Sleds."
5. Three point hydroplanes.
    (a) with two points forward.
    (b) with two points aft.
6. Hydrofoils.

In Fig. 151 is shown a modern stepless hydroplane. It will be noted that the bottom lines carry back so that the buttocks for at least the after third of the length are straight lines. In section, the "V" usually shows a slight amount of concavity — this amount depending

A MODERN STEPLESS HYDROPLANE

on the steepness of the angle of the "V." In a general way, the steeper the "V" angle, the greater the tendency for the boat to heel over to one side or the other at high speed unless this is corrected by concaving the "V." In very flat "V's" it is unnecessary to use much or even any concave.

This stepless type is exemplified in all modern high speed runabouts and in many high speed cruisers. For extreme speed, it is nowhere nearly as efficient as the other types but, since it is much closer to the normal boat form, it is a far better boat at low and medium speeds than the more radical hydroplanes. Its greatest fault at high speed, and by high speed is meant a speed-length ratio of approximately 10 or more, is a tendency to excessive porpoising. The farther aft the center of gravity in the stepless hydroplane, the more prone it is to porpoise. It is usually unsafe to have the center of gravity over 58 per cent of the water line length of the boat abaft the fore end of the water line. The center of gravity can go somewhat further aft if the last two or three feet of the bottom show a slight downward hook of perhaps a quarter of an inch. This hook, however, has other disadvantages and is not con-

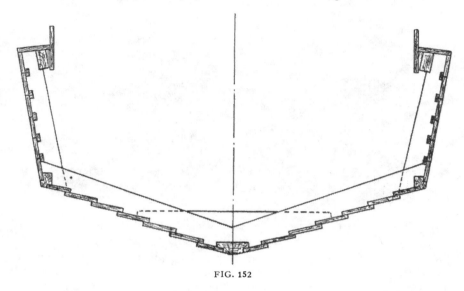

FIG. 152

ducive to safety in a sea nor to safety when making high speed turns and is, therefore, to be avoided if possible.

The exact action of the water on the bottom of a stepless hydroplane has been the subject of much discussion. It is conceded today that, in addition to upward pressures on the bottom, a great part of the area in contact with the water is subjected to a downward suction. It is for this reason that the stepless type is less efficient than any of the other forms in which an attempt is made to avoid these suctions.

Various forms of longitudinal steps have been used and have shown excellent results in many cases. These may be nothing more than the inverted laps of the planking created by starting the lapstrake bottom planking at the chine instead of the keel. A section of such a boat is shown in Fig. 152. These planks are usually wide at the bow and taper in toward the keel as they go aft.

By an easy transition from the stepless hydroplane, the multiple-step or "shingled" hydroplane is arrived at. In place of the smooth bottom of the stepless type, the bottom of the multiple-step hydroplane is broken up into a number of short planes with small steps behind them. For many years, it was considered that no step was effective unless it was more than two inches deep. Modern experiments, how-

A LONGITUDINALLY STEPPED HYDROPLANE

ever, have shown that a step even five-eighths of an inch deep is an effective step if properly vented so that air has a chance to get in behind the step. Many of the stepless hydroplanes which raced for the Gold Cup under the rules in vogue from 1923 to 1933 were altered to multiple-step planes by the addition of planes on the bottom applied in

FIG. 153

the form of wedges or "shingles." Usually four or five steps are used as shown in Fig. 153 which shows shingles applied to the stepless hydroplane of Fig. 151. The angles are slightly exaggerated for clarity. Points of shingles at keel and chine should touch straight lines as indicated and the aftermost plane should be the steepest. The multiple-step hydroplane, although not highly efficient, is, in general, next in seaworthiness to the stepless type. A slight change in plane angles, particularly the angles of the aftermost plane, has a marked effect on the running of the boat. Lowering the after step only one-eighth of an inch on *Delphine IV*, a multiple-step hydroplane, changes her from a porpoising hull to one which runs as smoothly as if on a track. *El Largarto*, a famous shingled hydroplane which won the Gold Cup year after year, is without doubt the most efficient hull of this type ever worked out.

The single step type of hydroplane, of which a typical example is shown in Fig. 154, is an efficient form when properly designed. Many hydroplane designers claim for it an efficiency greater than that

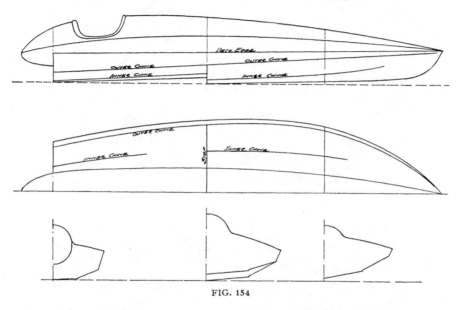

FIG. 154

of any other. This, however, is a matter of opinion but in all fairness it must be noted that the world's record holder, Sir Malcolm Campbell's *Bluebird*, and the famous *Miss Americas* of Gar Wood are all of the single step type.

In a single step hydroplane, the forward plane should have less angle than the after plane, the amount of difference in plane angle being approximately three-fourths to one and one-half degrees. A great variation is possible in the location of the step but it is customary to place it a little abaft amidships and to have the center of gravity of the whole boat about five per cent of its length abaft the step. This is not a hard and fast rule, only a guide to good practice.

Single step hydroplanes sometimes porpoise badly. They may throw the bow up into the air and then slam it down again, or may jump out at the stern. If the bow lifts badly in porpoising, the forward plane has either too steep an angle or the C.G. is too far aft. If there is a tendency for the stern to jump out and for the boat to come down on her forward plane, usually with a twisting motion, the angle of the after plane is too great or the C.G. is too far forward. It will be noted in the

"EL LAGARTO"

"DELPHINE IV"

section shown of the single step hydroplane that the sides are liberally flared from a comparatively flat and narrow bottom. At top speed, water does not touch these flared bottom sides and in turning they prevent water catching the chine and capsizing the boat as she skids sidewise. These non-trip chines are a feature of almost every high speed hydroplane of the single step type in use today.

The form of the single step hydroplane with the notch cut out amidships and the wide transom, is such a radical departure from the normal displacement form that it has excessive resistance at low and moderate speeds and must carry high power to drive it up to a point where it will plane. There is usually a sharply defined speed at which planing is definitely established. To reach this point may require full engine throttle but after planing has been attained the throttle may be partially closed and high speed maintained.

A form of hydroplane of unusual type which has many advantages for rough water work is the inverted "V" or patented Hickman "Sea

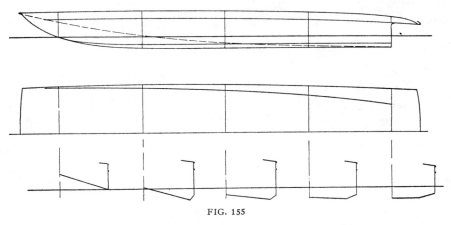

FIG. 155

Sled." This type is shown in Fig. 155. The inverted "V" runs almost the length of the boat, while the deck outline is practically a rectangle. This hull gathers under itself the bow wave and utilizes this wave to assist in lifting it. Although it has never been seen in any of the important competitions, many large boats of this form have been built for special purposes. The "Sea Sled" shown in the photograph is, from a weight-speed standpoint, one of the most efficient hulls ever built. On a weight of 34½ pounds to the horse power, the official speed was better than 46 statute miles an hour.

In a following sea, the action of the "Sea Sled" is excellent. Due to the form of the bow, there is little tendency to yaw as the stern lifts. The inverted "V" is a true planing hull requiring a light and powerful engine and it has a decided planing point.

The latest hydroplane development is often termed a "three-point" hydroplane. The basic idea of the three-point hydroplane is to form the bottom so that there are three separate planes, each working in water which has not been disturbed by any plane ahead of it. This tends to give great efficiency, since planes which run in disturbed water have far less lift for a given amount of wetted surface than planes passing through undisturbed water. The general idea of a three-point hydroplane having two points of support forward and one point aft is shown in Fig. 156. In the 225 cubic inch hydroplane

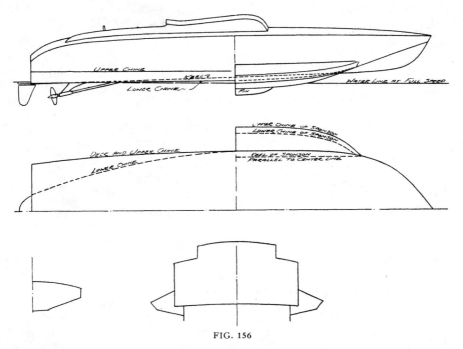

FIG. 156

class, all successful hulls today are of this type. Official records of better than 72 miles an hour, with a motor of not over 225 cubic inches piston displacement developing 165 horse power, have been made.

For a single screw hydroplane, the two points of support, which may be considered little "V" bottom hulls, should be forward. If the boat has twin screws, some advantage is to be gained by making the two supporting planes at the after end of the boat with the single plane forward. In order to get stability, the C.G. of the whole boat should be not too far abaft the two supporting surfaces when they are forward, and not too far forward of them if they are aft. In no case should the weight be near the single center plane.

Since each plane works in undisturbed water, the efficiency of the three point hydroplane is considerably better than that of any other form. It is this form which promises to reach even greater speeds than the *Bluebird*, Sir Malcolm Campbell's single step hydroplane, which has an official record of approximately 130 miles an hour.

The sixth class of gliding hull is that in which the weight of the boat is supported by small submerged hydrofoils shaped on the principle of the wing of an airplane. The proponents of the hydrofoil idea claim for this type of hydroplane an efficiency better than can be found with any other form. This, however, is open to discussion as the "Sea Sled"

A HICKMAN "SEA SLED"

and the three-point hydroplane have shown efficiencies as high as have been reported for any hydrofoils. Since the main supporting surfaces of the hydrofoil are below the water, it has no inherent stability of its own. Stability at speed must be gained by having the foils arranged in series, one above the other, so that at the highest speeds only the smallest and lowest foils are under the water and they cannot have an area sufficient to lift to the surface; but stability is none too good at best.

Several partially successful craft have been built on the hydrofoil idea but it does not lend itself to use of the under-water propeller. The greatest hydrofoil successes have been attained with the use of an air propeller and air propelled craft are beyond the scope of this chapter. No hydrofoil has as yet ever competed successfully with the other types in any important competitions.

From the foregoing, it will be noted that it is a most difficult and an almost impossible problem to calculate the resistance of any one of the various forms of hydroplanes by any scientific method. An approximation to the probable speed of a well designed hydroplane may be obtained from an empirical formula which is as follows:

Divide the total actual running weight of boat, complete with crew, in pounds, by the actual horse power of the motor. Obtain the square root of this weight per horse power and then divide a coefficient, values of which will be given later, by the square root of the weight per horse

"TOPS II", A THREE-POINT HYDROPLANE

power to obtain the probable speed in statute miles an hour. Expressed as a formula: $\dfrac{C}{\sqrt{\dfrac{W}{P}}} = S$

Where C is the coefficient, W is the total weight in pounds of the boat in running condition, P the actual brake horse power, and S the speed in statute miles per hour. This coefficient has the following approximate values:

For ordinary stepless hydroplanes, such as high speed runabouts, C equals 180 to 185. For multiple-step or shingled hydroplanes, C equals 190 to 205. Single step hydroplanes of good design, C equals 210. For "Sea Sleds," C is 220 for small sleds to 270 for the largest and most efficient. For small three-point hydroplanes, C is about 240 to 250. Values for hydrofoils cannot be included in this formula as the data are not available.

This formula makes two assumptions: that the resistance of any one of these types varies directly as the displacement and directly as the speed. This may appear to be faulty mathematics and to some extent it is, as it does give an advantage to large boats. The error, however, is much less than appears at first glance as it takes care of the resistance created by shafts, struts, rudders, skid fins, and other underwater details which cause the resistance of the actual hydroplane to be quite different from that calculated from model tests on a hull without these appendages. At the present time, when no high speed tank for testing hydroplanes is available, this simple formula is, in practice, as nearly correct as the most complicated calculations.

CHAPTER XVII

SCREW PROPELLERS

THE following is a list of the principal characteristics of propellers and the symbols by which they are denoted, together with some of the ratios by which their relationships are indicated:

$$d = \text{diameter in feet} = \frac{101.3\ V_w}{a\ R(1\text{-}s)}$$

$$p = \text{pitch of face in feet} = d\ a = \frac{101.3\ V_w}{R(1\text{-}s)}$$

$$a = \text{pitch ratio} = \frac{p}{d}$$

$$M.W.R. = \text{mean width ratio} = \frac{\text{average width of blade}}{d}$$

$a.\ t.\ r. = \text{axial thickness ratio}$

$V = \text{speed of vessel in knots}$

$V_w = \text{speed of screw through water in knots} = V(1\text{-}w)$

$$s = \text{real slip ratio} = \frac{p\ R - 101.3\ V_w}{p\ R}$$

$$s' = \text{apparent slip ratio} = \frac{p\ R - 101.3\ V}{p\ R}$$

$$w = \text{wake factor} = 1 - \frac{V_w}{V}$$

$R = \text{revolutions per min.}$

$$T = \text{thrust} = \frac{R_t}{1 - t.}$$

$$t = \text{thrust deduction coefficient} = \frac{1 - R_t.}{T}$$

$$\text{hull efficiency} = \frac{1 - t}{1 - w}\ (\text{generally assumed} = 1)$$

P = Brake hp.
D = delivered hp. = P — bearing losses
$T.P.$ = thrust hp. = $T \times .00307\ V$
E_t = effective hp. = $R_t \times .00307\ V$

Propeller efficiency = $\dfrac{T.P.}{D.}$

Propulsive coefficient = $\dfrac{E_t}{P}$

DIAMETER

The diameter is the most important feature of a propeller. A slight change in diameter has more effect on power absorption than a considerable change in pitch or blade area. Where the ideal diameter is greater than is desirable for practical considerations, it may be reduced, necessarily with some loss of efficiency, by varying blade area and shape, pitch and the number of blades to keep power absorption constant. Fig. 157 gives data necessary for ready calculation of diameter and pitch by Taylor's method, the use of which will be described later.

PROPELLER CHARACTERISTICS

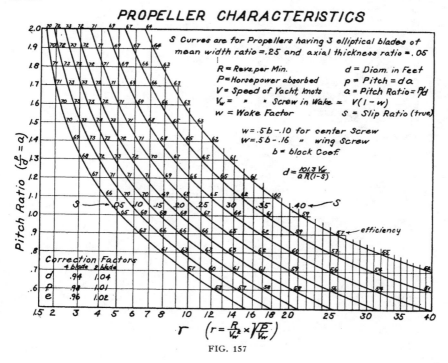

FIG. 157

PITCH

The pitch used in calculation is that of the after or driving face of the blade and is known as nominal or face pitch. It is the distance the propeller would advance in one revolution if it had no thickness and if there were no slip. The nominal pitch is not the actual pitch, since the blade has thickness and the suction of the back of the blade exerts an influence so that the combined effect of the face and back is to give a true pitch greater than that of the face pitch.

The true pitch is the distance the actual propeller advances in one revolution when not exerting any thrust. There is always some thrust

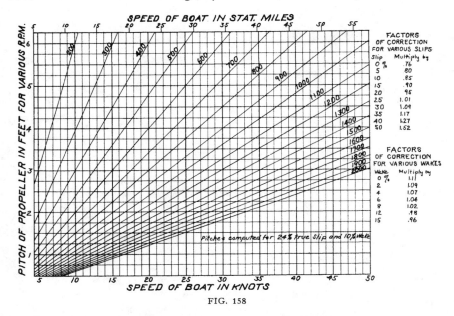

FIG. 158

with a propeller advancing at such speed and revolutions that the slip ratio theoretically is 0. As a rule, the greater the axial thickness ratio, the greater the amount the true pitch exceeds the nominal pitch. Propellers are nearly always designed to be true screws, that is, the pitch is constant at all distances from the axis and the blades are flat so that the pitch is the same at the forward and the after edges of the blade. The true screw is in practically universal use for reasons of necessity of standardization of model test data, and for simplicity of design and construction.

If all portions of the propeller were moving through the water at equal speed, the true screw would accelerate water sternward at equal rate, but the boat drags water along with her, more or less at different speeds at different distances from the hull. This is called the wake. In a single screw boat with shaft emerging from the sternpost, the wake is greater near the hub than farther out, so the pitch should be less near

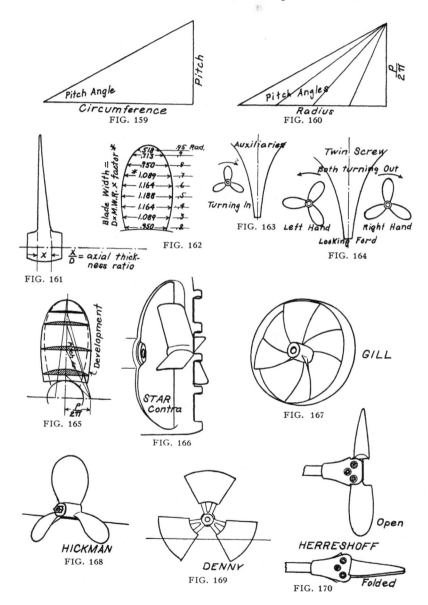

FIG. 159

FIG. 160

FIG. 161

FIG. 162

FIG. 163   Left Hand   Right Hand

FIG. 164

FIG. 165

FIG. 166

FIG. 167   GILL

HICKMAN
FIG. 168

DENNY
FIG. 169

HERRESHOFF
FIG. 170

the hub to give equal slip at all distances out from the axis. In such cases, pitch increasing from hub to tip will give greater efficiency than a true screw.

Fig. 158 gives pitches in feet for various speeds and revolutions.

PITCH RATIO

The ratio of the pitch to the diameter is known as the pitch ratio. Practicable pitch ratios range from .5 to 2.0; below and above these limits the efficiency is very low.

PITCH ANGLE

The blade angle at the tip will be found, as in Fig. 159, by laying off to scale the circumference as a straight line and the pitch perpendicular

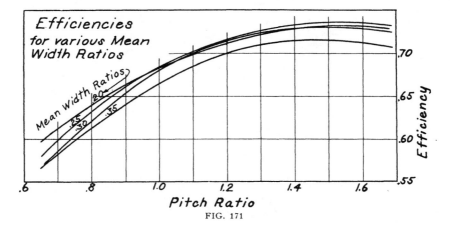

FIG. 171

to it. It is more convenient to lay off radius than circumference, so if we lay off $\frac{p}{2\pi}$ instead of pitch, and radius instead of circumference, we get the same angle. For a true screw the pitch is the same at all distances from the center, so pitch angles at various radii are as indicated in Fig. 160.

MEAN WIDTH RATIO

The mean width ratio is the ratio of the average width of the blade to the diameter of the screw. The average width is equal to the area of the blade divided by its length from hub to tip. This ratio is a better index of blade area than ratio of developed area to disk area, which is often used. One might suppose that thrust would be proportional to blade area and that if a blade of, say, .25 mean width ratio were increased 10 per cent, the thrust would also be increased 10 per cent. This is not the case, however. The thrust increases little with increase in area made in this way; indeed, at low pitch ratios the thrust is less for a wide blade than for a narrow blade.

This is, to some extent at least, due to the greater blade thickness in proportion to width of the narrow blade, which gives a virtual pitch greater than the nominal or face pitch as already described.

In general, narrow blades are more efficient. Fig. 171 shows efficiencies of four mean width ratios for various pitch ratios.

The reason for the lesser efficiency of wide blades is that the forward portion of the blade is entering solid water that has not been set in motion by friction with the blade, and the portion along the leading edge performs relatively more work on this account. Narrow aeroplane wings, rudders and other pressure surfaces are more efficient than wide ones for this same reason.

.25 is a good all-round mean width ratio to use. .20 is a good value for non-feathering blades which are to stand vertical behind the stern post of auxiliaries when sailing. Very narrow blades are not practical as it is impossible to make them strong enough without getting the thickness too great for efficiency.

Wide blades, up to .40 mean width ratio, may be needed where there is danger of cavitation and it is necessary to keep the thrust per square inch of area down to a practicable amount.

AXIAL THICKNESS RATIO

The axial thickness ratio is the ratio of the thickness the blade would have if face and back were extended to the axis, divided by the diameter, as shown in Fig. 161.

The axial thickness ratio is usually about .05 for adequate strength and this is also the best axial thickness for efficiency. Narrow blades are more efficient with a less amount than this and wide blades with a greater amount. The axial thickness ratio should be about one-fifth of the mean width ratio for best efficiency.

WAKE

The water in which the screw is working is not at rest but is travelling with the vessel, through the action of skin friction, at an average speed expressed as a fraction w, known as the wake factor, of the speed of the vessel V.

The screw, therefore, is travelling through the water at a less speed than the vessel and this speed is denoted by V_w which is equal to

$V(1-w)$. As already explained, the wake is not uniform all over the area acted on by the propeller and this is one cause of propeller vibration. The wake factor w varies from almost 0, in the case of high speed craft with flat sterns and no deadwood, to .20 or more for slow craft with deadwood and large displacement. A means of approximating w for straight power craft of moderate speed and proportions is indicated in Fig. 157.

NUMBER OF BLADES

Efficiency is, roughly, inversely proportional to the number of blades, but the difference is small. The two-bladed propeller is little used except for auxiliaries with propellers operating in an aperture in the deadwood where there is an advantage in locking the propellers in a vertical position when sailing. A propeller blade exerts a varying thrust at different stages of its rotation, due to the influence of strut, deadwood or hull, which causes vibration, and this vibration is lessened by increasing the number of blades. Four is the greatest number of blades that has been found practicable and four are resorted to only where it is desirable to keep the diameter as small as possible, three being the standard for usual conditions.

SHAPE OF BLADES

The shape of the blade is not a matter of much importance and on the score of efficiency the elliptical blade is as good as any. Where it is desired to keep the diameter small, the blade may be expanded at the tip, making a broad tipped blade. Area added at the tip will increase the thrust about in proportion to increase in area, but with some loss of efficiency. Fig. 162 gives factors for constructing an elliptical blade. The product of the diameter, mean width ratio and these factors give the width at various fractions of the diameter from the center.

SLIP

Since water yields under pressure, the propeller necessarily operates with some slip. A propeller large enough to operate with very little slip would lose in efficiency (on account of excessive friction) much more than it would save from small slip loss. It will be noted from Fig. 157 that maximum efficiencies are obtained with 20 to 25 per cent slip.

The slip used in propeller calculations is the true slip based on the speed of the propeller through the wake. When calculated with reference to the speed of the boat, it is called the apparent slip.

A common difficulty with gasoline engine propelled craft, especially auxiliaries, is the excessive slip necessitated by the engine revolutions being too high for the speed of the boat. The loss in efficiency is often offset, however, for practical considerations by the saving in space and weight of the propelling outfit on account of its high revolutions.

HULL EFFICIENCY

Power is gained from the wake on account of the propeller working in water already in forward motion, but this gain is about counterbalanced by the thrust loss, due to the suction of the propeller on the hull. Both wake and thrust are difficult of determination, but many tests have shown that w and t are approximately equal for normal hulls so that the hull efficiency, which is $\dfrac{1-w}{1-t}$, may be taken as unity.

DESIGN

The law of comparison applies to propeller resistance as well as to hull resistance and on this account it is possible to design full sized propellers by the use of data acquired from tests made in the towing tank on small model propellers. A great deal of such data has been accumulated by Froude, Admiral Taylor, Professor Durand and others, which afford us sufficient information for the accurate designing of propellers for almost all purposes. Professor Durand used four-bladed models 12″ in diameter, and Admiral Taylor used 16″ three-bladed models. These tests were run for a wide range of pitch ratios, slips, mean width ratios and axial thickness ratios.

The law of comparison applied to propellers means that the torque and thrust of a full sized propeller are to the torque and thrust of the model propeller as the cube of the ratio of their diameters. This stipulates that the slip is the same for both and the ratio of their speeds is proportional to the square root of the ratio of their diameters.

The law of comparison works out well in practice and proper proportions of large propellers may be accurately determined by the use of the extensive data worked up from model tests. This is extremely fortunate, as the accumulation of any amount of coherent data from large screws, involving all sorts of combinations of power, speed and revolutions, is out of the question.

I have consolidated Admiral Taylor's propeller design data in Fig. 157 in a form which makes it practical for use in determining proportions of propellers of all sizes within the limitations of current practice. The usual problem involves a certain brake horse power, speed of vessel and a certain number of revolutions of the engine.

Knowing V and estimating w, we calculate V_w. We then have all factors necessary for calculating the quantity r. Knowing r, we read directly above it to the slip curve showing the best efficiency. Efficiencies are indicated in small figures along each curve. From the intersection of this vertical with the selected slip curve, we follow horizontally to the left and read the corresponding pitch ratio. Having thus found the slip, efficiency and pitch ratio, we substitute these quantities in the second formula and compute the diameter in feet. The pitch, then, is equal to the diameter times the pitch ratio. The pitch and diameter obtained as above are for three-bladed screws. If a two-bladed or four-bladed propeller is desired, the diameters, pitch and efficiencies may be found by multiplying by the factors given in the lower left hand corner of the diagram.

If the diameter as computed for maximum efficiency proves too great for local conditions, such as size of screw aperture or hull clearance, a smaller diameter may be computed by selecting a higher slip than that first chosen, and substituting the corresponding pitch ratio in the formula for diameter. The following example will illustrate the use of this figure:

Given

$$V = 10 \qquad R = 400 \qquad P = 50 \qquad w = .10$$
$$\text{then } V_w = 10 \times .90 = 9,$$

using Fig. 157
$$r = \frac{400}{81} \times \sqrt{\frac{50}{9}} = 11.7,$$

from the curves, for best efficiency

$$s = .25 \qquad \text{pitch ratio} = 1.0 \qquad e = .67$$
$$\text{then } d = \frac{101.3 \times 9}{1 \times 400 \times (1 - .25)} = 3.04 \text{ ft.} = p$$

If diameter is to be kept as small as possible, use top curve.

$$s = .40 \qquad a = 1.42 \qquad e = .62$$

$$\text{then } d = \frac{101.3 \times 9}{1.42 \times 400 \times (1 - .40)} = 2.67 \text{ ft.}$$

and $p = 2.67 \times 1.42 = 3.80$ ft.
For 4 blades $d = 2.67 \times .94 = 2.51$ ft.
and $p = 3.80 \times .98 = 3.72$ ft.

POSITION OF PROPELLER

The highest efficiency is obtained by locating the propeller as far aft as is practicable, and as near the surface as possible without breaking water. In so doing, the disturbing influences of wake, thrust loss and stream line disturbance are minimized.

In the case of auxiliaries, a special problem is involved. If the propeller is located on the center line, it must be low down to avoid disturbing the stream lines of the run or weakening the rudder by taking a large part of the aperture out of it. Locating the propeller on one side of the stern post (see Fig. 163) has several decided advantages. With the propeller in this location, the form of a pure sailing vessel is preserved; there is no aperture to cause eddies and interfere with steering and, if the wheel turns inboard, a neutral rudder is had when under power.

Most marine engines turn right-handed, that is, revolve like the

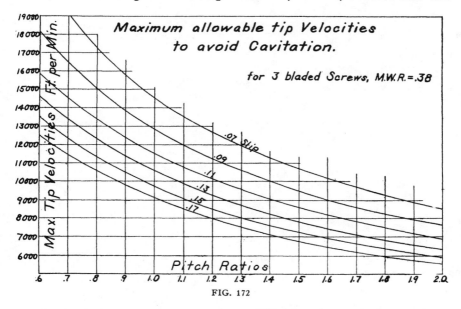

FIG. 172

hands of a clock when viewed from aft. With a single screw located on the center line, there is a tendency for the bow of the boat to go to port. This must be counteracted by the rudder. With twin screws, the almost universal practice is for the screws to revolve in opposite directions and outboard (see Fig. 164), that is, the blades turn outboard from the top position — right-handed for the starboard screw and left-hand for the port. In this way, neutral rudder is had with both screws turning while with one stopped the maximum turning moment is exerted by the other. With a single screw on one side turning inboard, the tendency of the screw to throw the stern to one side is counteracted by the eccentric direction of the application of its thrust on the hull.

CAVITATION

If the screw is turned at too high revolutions, the water is no longer able to remain in solid contact with the blades at all points. This is known as cavitation and is to be avoided or allowed for in proportioning the blades as it greatly impairs the efficiency, especially of narrow blades.

Cavitation is worst near the leading edge of the driving face so the after portion of the blade does the work when there is moderate cavitation, hence the value of the wide blade. Wide, thin blades also permit the leading edge to be made thin and sharp, which is desirable.

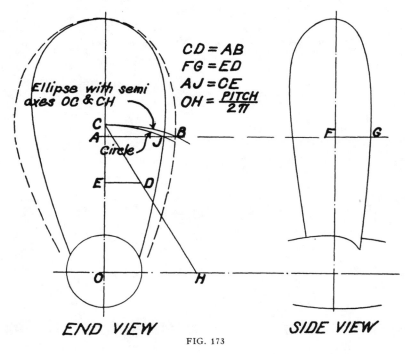

FIG. 173

To avoid cavitation, the tip speed should be kept within certain limits which vary widely with the slip and pitch ratio. These limits, as indicated by Barnaby, are given by the curves in Fig. 172. Tip speed decreases, of course, with decrease of diameter, revolutions being constant; to reduce diameter to the minimum, one may employ four broad blades with wide tips.

LAYOUT

The layout of a conventional true screw is shown in Fig. 165. The blade sections are circular on the back, and the angle each section makes with a plane perpendicular to the shaft (pitch angle) is indicated by the angle each slanting line makes with the vertical axis. The manner in which the end and side projections are obtained from the developed area are shown in Fig. 173.

For efficiency, the edges of the blade should be as sharp as possible.

PROGRESSIVE SPEED TRIALS

Fig. 174 shows the results of a progressive speed trial on a small displacement type speed launch designed many years ago by N. G.

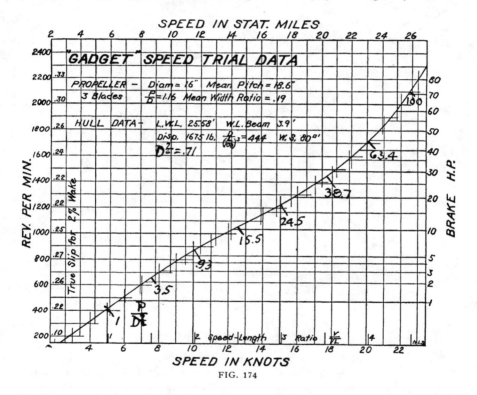

FIG. 174

Herreshoff. Several larger and successful speed boats were also built from this design. Speed trials should be made whenever possible, that is, the boat should be run over a measured course at various revolutions and the speed and power figured. In this case we have a single curve showing revolutions, slip, and horse power for various speeds up to 27 miles per hour. It is particularly interesting to note how the slip varies at various revolutions.

SPECIAL TYPES

Since the first application of the screw to marine propulsion, this device — like many other important discoveries — has been the subject of countless inventions intended to increase the speed of vessels. When we consider that the efficiency of the simple true screw propeller may run as high as 75 per cent, it is plain that great improvement is impossible. It does not follow, however, that slight but worthwhile improvements are impossible, particularly for special conditions such as shoal draft. A description of a few modern types of screws of special merit follows.

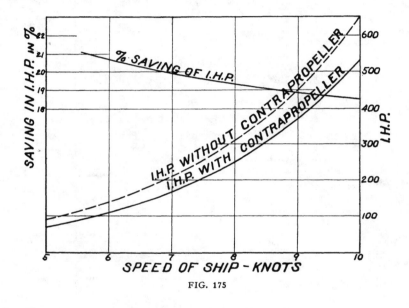

FIG. 175

THE STAR CONTRAPROPELLER

This device consists of fixed blades just abaft the propeller, set at an angle to extract a forward thrust from the rotation of the propeller jet. Savings in power of from 8 to 18 per cent are realized in practice over the best possible propeller. Fig. 175 shows the results of trials of the 202 foot cargo steamer *Havmöy* with and without the contrapropeller. Fig. 176 shows the reaction of the propeller jet on the contrapropeller, and Fig. 166 shows the appearance of the device. It has been in use abroad for a number of years, and the large number of vessels equipped with it testify to its genuine merit. In addition to raising the efficiency of propulsion, it improves steering and cuts down vibration.

The device is also installed ahead of the propeller to start the water rotating in a direction contrary to that of the screw. This results in a further improvement in efficiency equal to that of the contrapropeller abaft the screw. Where a propeller shaft is supported by a V-strut, the

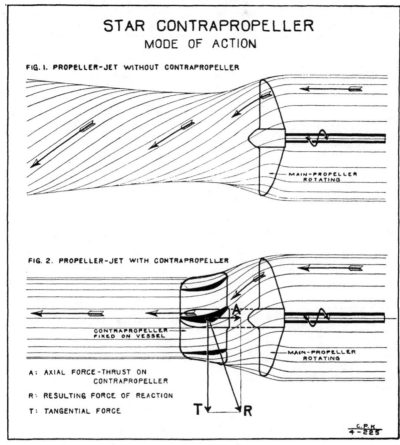

FIG. 176

arms of the strut and a vertical guard (cast with the strut) below the axis may all be formed with a twist, making a three-bladed contra-propeller.

THE GILL SHROUDED PROPELLER

The Gill propeller, shown in Fig. 167, is fitted with an annular ring or shroud which contracts in nozzle shape from the forward to the after end of the propeller, and the face pitch of the blades increases axially in the same direction, producing a regular acceleration of the thrust stream. Since the stream is directed straight aft, this propeller is particularly effective where there is oblique flow to the propeller and where the propeller must be installed close to the surface.

Its other advantages are quick acceleration, 10 to 15 per cent smaller diameter, better steering, lessened vibration and improved efficiency.

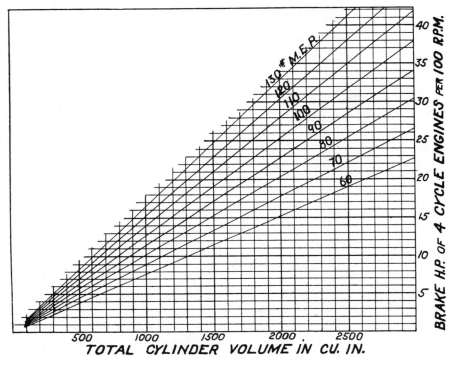

FIG. 177

HICKMAN SURFACE PROPELLERS

The obvious advantages of surface propellers are shoal draft, weedlessness, location well aft, great maneuverability if twin wheels are separately driven, stern bearings above water, high efficiency due to horizontal shaft line, and elimination of all parasite resistance of shaft, struts and propeller hub. (See Fig. 168.) To offset these is cavitation from disturbed surface water and splash. These propellers must necessarily be of greater diameter than the equivalent submerged propeller at same revolutions (30 to 50 per cent greater), and must rotate at less speed than the maximum possible with a submerged propeller to keep the pitch ratio and the tip speed within reasonable limits. The immersion is 30 to 40 per cent of the diameter. They are usually fitted in pairs, rotating in opposite directions to neutralize side thrust, though a single screw may be used if placed in correct relation to the rudder or rudders.

THE DENNY VANE WHEEL

These wheels, patented by William Denny & Bros., Ltd., of Dumbarton, Scotland, (Fig. 169), are similar to the Hickman Surface Pro-

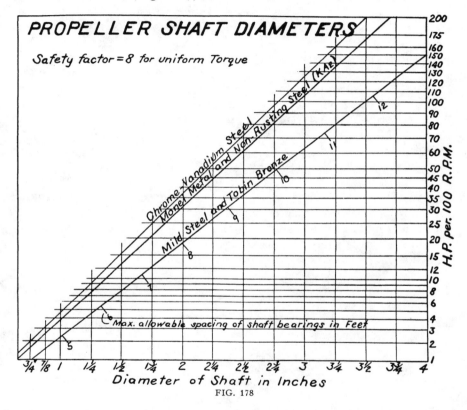

FIG. 178

pellers, and possess the same advantages. They are a built-up wheel of large diameter, operating at low revolutions. They are adapted to houseboats and shoal draft moderate speed craft of many types.

FEATHERING PROPELLERS

The use of a propeller which will offer little resistance when sailing is important in the case of auxiliaries. One of the best of such propellers is the Herreshoff folding propeller shown in Fig. 170. When sailing, a catch is released which permits the blades to fold back by action of the pressure of the water on them, and in this position they cause the least possible amount of drag. When the engine is started up, the blades fly into position through centrifugal action and are locked in position by the catch, so they cannot fold back if engine is reversed.

The Thompson feathering propeller has its blades pivoted on the hub and the center of pressure of the blade abaft the axis. When sailing, the blades automatically swing into the fore and aft plane and when engine is started they swing into driving position, being held from swinging further by a stop. The driving face is the pressure face in both ahead and astern positions, which makes the reversing efficient. This wheel must be used with a reverse gear if reversing is desired.

Fig. 177 is useful for approximating the horse power of 4-cycle gasoline engines. Determine the cylinder volume, which is piston area times stroke times number of cylinders, and read the hp. per 100 r.p.m. on the probable M.E.P. (mean effective pressure) curve. The M.E.P. is about 80 or 90 pounds per square inch in the ordinary engine, though it may run as high as 130 in refined, expensive engines. The power per 100 r.p.m., multiplied by the number of hundred revolutions, gives the total b.hp.

Fig. 178 is useful for determining the diameter of propeller shaft. The number of hp. transmitted per 100 r.p.m., times the number of hundred revolutions, gives the total horse power that may safely be transmitted by a shaft of a given size.

CHAPTER XVIII

YACHT MODELS

THERE is a widespread and growing interest in scale models of all sorts of vessels and especially in yacht models, for in the miniature one can take in the boat as an entity. In the case of the real vessel, the observer is hampered in doing so by the difficulties arising from the relative disparity in size between yacht and observer, and from the partial concealment of the hull by cradle, staging, other boats, buildings, etc., if she is on shore, or by the water if she is afloat.

For purposes of discussion, yacht models may be divided into a number of classes, according to the use to which the model is to be put, such as designers' models, builders' models, tank test models, ornamental models and sailing models.

DESIGNERS' MODELS

The scale model (usually a half model) furnishes a most useful means of studying the appearance, performance and general characteristics of the full sized yacht. It is obvious that, even to the expert designer, plans cannot convey as completely visualized an impression of the form of the vessel as does the model, carefully made to those plans; to the layman, the model means ever so much more than the "lines."

Formerly many rule-of-thumb designers, especially of commercial sailing vessels, first made the half model entirely by eye, but to a certain convenient scale, and then made the lines from the model by sawing it across on stations and marking around the sections on paper. Sometimes the model was made up of horizontal lifts pegged together. When the model was all faired, the lifts were taken apart and used as templates for constructing water lines on paper from which measurements were taken for laying down the full sized vessel.

The modern designer generally makes small use of the model but it may be advantageous in some instances to begin drawing the lines and bring them and the model along together, determining displacement, centers of buoyancy, etc., on the lines, and refining the form on the model. N. G. Herreshoff made much use of half models when designing his famous racing yachts. He even took offsets from the model, using a special machine of his own invention.

In making a designers' half model, some straight grained, easy working wood is used. Soft pine is the favorite, though basswood, Spanish cedar, mahogany and California redwood are suitable. A nice combination is pine topsides with redwood bottom, mounted on a black backboard; the contrast between topsides and bottom is useful. Lifts are marked off from the lines, using buttocks for vertical lifts, or water lines for horizontal lifts. These should be sawed out a trifle "strong" and planed to the exact thickness. After the lifts are glued together, the corners between the lifts form a guide to which the model may be roughly shaped. Templates, made to the transverse sections, should be used in working the model down to the final surface. If the wood is nice and clear, as it should be for the model, it is easier and more satisfactory to finish it bright than to attempt to enamel or paint it.

BUILDERS' MODELS

Where a vessel is built with a metal skin it is necessary to make an accurate half model on which to lay out the strakes of plating, the frames, butt straps, ports, etc. The model is made with $\frac{3}{8}''$ or so of extra thickness the other side of the central longitudinal plane to avoid sharp edges on the profile. The model is painted white, as a rule, and the plates, etc., are laid out on it in ink. The true shape of the plates is thus indicated and the stock can be ordered from the model.

TANK TEST MODELS

These are, of course, full models and are made to as large a scale as practicable to get dependable results. The manner of obtaining data on the resistance of the full sized vessel from that of the model is outlined in Chapter XV. There is great value in tank experiments for large and important vessels, but the expense and the time required are often prohibitive for yacht work. One can, however, do some interesting and instructive work in testing one model against another by towing them in still water, side by side, from the ends of a beam, thus finding the comparative resistances of the two. This towing should be done ahead of a launch, from a bowsprit or projecting spar, to avoid the wake of the launch. The performance of the models at various speeds can be studied closely and much of value learned.

Another comparatively cheap substitute for the towing tank is the whirling arm. On a platform in a small island in the center of a small pond a pivoted arm is erected, thirty or forty feet long, or even more, which may be rotated by a motor at any desired number of revolutions per minute. Knowing the length of arm, the number of r.p.m. for any speed of model is readily computed. The tow line should be attached to the bow of the model at a height and at an angle approximating the direction of the thrust from the propeller. The tow line may be led to sensitive scales at the inboard end of the arm where the exact resistance may be read. The trim and flotation of the model may be recorded by photography, using a camera mounted part way out on the arm.

This apparatus is particularly useful for studying hydroplane forms.

ORNAMENTAL MODELS

Among these I include nicely finished half models for decoration of home, club or office; rigged yacht models and the popular ship model. Many ship models are highly inaccurate and abhorrent to the trained eye. Although two-sided models may never be placed in the water, they should be hollowed for lightness and to prevent checking. The methods followed in making the hull are practically the same as those described under Sailing Models. The success of the ornamental model depends on scale accuracy, carried through to the smallest fitting. It is a waste of time and material to start out to make a model by eye.

SAILING MODELS

These include models of particular vessels or types not intended for racing, and racing models built to a special measurement rule and intended for competition. Scale models of larger vessels do not perform well. Their stability is insufficient, for the wind velocity is not scaled down and stability varies as fourth power of a lineal dimension, whereas heeling effect varies as the cube. Therefore scale models have relatively much less stability.

Racing models are an entirely different proposition. They are built to a given rating rule to give maximum speed under that rule, and are necessarily different in proportions and details of form and rig from any full sized yacht built to the same rule. Figs. 179, 180 and 181 give the principal characteristics of the three most popular classes under which models are raced today. Class B (Fig. 179) is built to the Universal Rule just as are large yachts, but with a limit of 46″ rating, and different limits on displacement and draft. The displacement varies directly with the length, producing a constant sail area for normal displacement, and the draft limit produces greater draft in proportion. Yachts built under this rule run 55″ to 60″ L.W.L. for the best results, making quite a large, heavy model.

Contestants in the International Races are models built to Class A, which are Six-Metre models on the scale of 2″ to 1′. Fig. 180 gives the rule for this class and the sail area and draft for normal displacement, with no quarter beam or other penalty. The rating formula is as follows:

$$\frac{L+\sqrt{S}}{4}+\frac{L\sqrt{S}}{12\sqrt[3]{D}}=\text{Rating}$$

Where L = Load Water Line length in inches plus ½ any excess in quarter beam measurement.

$\sqrt{S}$ = The square root of the total sail area in square inches measured in accordance with I.R.Y.U. regulations.

$\sqrt[3]{D}$ = The cube root of the displacement of the model in cubic

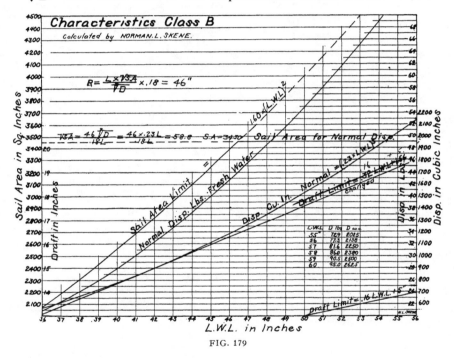

FIG. 179

inches, in full racing trim with largest suit of sails, including spinnaker or other running sail.

Load water line length ($L.W.L.$) is the distance in a straight line between the points farthest forward and aft in the plane of flotation in full racing trim.

Load water line beam is the extreme breadth in the plane of flotation.

Quarter beam measurement is the quarter beam length ($Q.B.L.$) measured in a line parallel with the middle fore and aft vertical plane (or center line) at a distance from it equal to one-quarter of the load water line beam, and one-tenth of this breadth (*L.W.L.* beam) above *L.W.L.*

Excess of quarter beam measurement is the amount by which the *Q.B.L.* exceeds the length allowed without penalty. This length is equal to a percentage of the *L.W.L.* length found by subtracting the square root of half the load water length in inches from 100. (Percentage $= 100 - \sqrt{\frac{1}{2}L.W.L.}$.)

Displacement in cubic inches is the weight of the model in pounds avoirdupois divided by .037.

STARTING "PRINCE CHARMING" AND "LADY NELL"

Limits and Penalties: (*a*) There shall be no limit to the displacement of models, but the cube root of the displacement ($\sqrt[3]{D}$) as used in the measurement formula shall never exceed one-fifth of *L.W.L.* (in inches) $+1$, but in the event of it being less than one-fifth of *L.W.L.* (in inches) $+.4$, then an amount equal to the deficit shall be deducted from the actual cube root for use in the measurement formula.

(*b*) The maximum draft shall not exceed *L.W.L.* $\times .16 + 3.5$ inches. Any excess in draft shall be multiplied by 3 and added to rating.

(*c*) Average freeboard (taken at center and at forward and after ends of *L.W.L.* to top of deck below rail) shall not be less than $\sqrt[3]{D} \times .28 + 1$. Sheer shall be a fair continuous curve. Any deficit in freeboard shall be added to rating.

(*d*) Height of sail plan above deck shall not exceed 85.3 inches. Any excess shall be added to rating.

(*e*) No hollows are allowed in the surface of the hull between the L.W.L. and sheer line. Stem and stern profiles must be fair easy curves.

(*f*) Any local concave jog or notch in the plane of measurement at either end of the L.W.L. shall be bridged by a straight line, and the L.W.L. taken to the intersection of such lines with the established

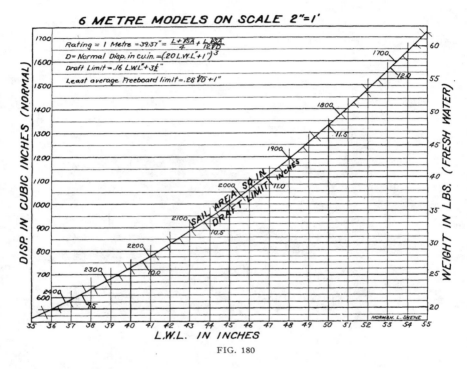

FIG. 180

L.W.L. plane. Any concavity in the stem line shall be bridged by a straight line equal to one-third of the greatest load water line beam, placed equally above and below L.W.L.

(*g*) The round of deck beams must not exceed one-twelfth of an inch for every two inches of the beam.

(*h*) Centerboards, lee boards and bilgeboards are prohibited.

(*i*) The number of battens in the leach of the mainsail shall not exceed four and they shall be equally spaced. Intermediate battens shall not be longer than 7.87 inches, and top and bottom battens 5.90 inches. Headsticks to a triangular mainsail shall not exceed .98 of an inch, and any sail with a longer headstick shall be measured as a gaff sail. The headstick of a spinnaker must not exceed one-twentieth of the measured length of the spinnaker boom.

(j) Any increase of sail area obtained by the use of intentionally bent masts and spars will be measured as a bow and included in the sail area.

Booms. In the case of booms other than those round in section, half the depth in excess of 1 inch shall be added to the sail area.

"BLUE CHIP"
U. S. Challenger, 1936

"LADY NELL"
World's Champion, 1936

Masts. If the section of a mast is greater in a fore and aft direction than athwartships, the difference shall be included in the area of the mainsail, and of topsail.

(k) Models must always sail with masts and spars as measured.

(l) Hollow masts and spars are allowed.

(m) There are no restrictions as to scantlings or materials.

(n) Models must be measured in salt water. The weight of sea water to be taken as 64 lbs. to the cubic foot.

The maximum height of the fore triangle shall not exceed 64 inches (i.e., 75 per cent of the height of sail plan allowed). The maximum length of luff and leach of the spinnaker is $0.8\sqrt{I^2+J^2}+16.4$ inches. I is the hoist of the headsails and J the base of the fore triangle. The maximum width is $2\times J$.

Class A models are usually about 74″ to 80″ in length over all, 14″ to 15″ beam, 48″ to 52″ length on the water line and from 45 to 53 pounds displacement.

As in full size yachts, it pays to design the model so that there will be no penalties. Under this rule, L is measured as under our Universal Rule and not under the International Rule as is the case with full size Six-Metre yachts. Yachts built in this class run about 4 feet on *L.W.L.*

Models are also built to the R class, built one-eighth the size of full size yachts, and with no changes in restrictions or methods of measurement. This makes a model of convenient size, running near the top of the sizes indicated on the curve, Fig. 181. A model 45″ on *L.W.L.*

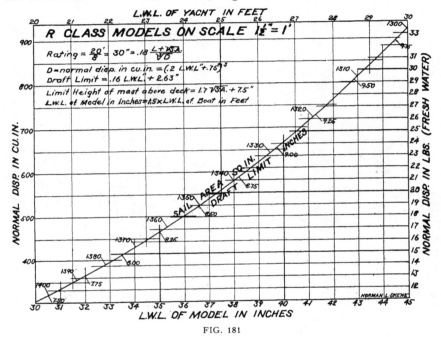

FIG. 181

will probably do better, on the average, than one 40″, which would correspond to about the largest full size yacht which can win under ordinary conditions. Remember that many times the model sails under what would be gale conditions for the full size yacht, and the big hull with its extra buoyancy and stability has the advantage there.

Another popular class in American waters is known as the Marblehead 50–800. In this class, the over all length of the hull shall not be greater than 50″ and the sail area shall not exceed 800 square inches. One quarter of an inch in excess or less than the 50-inch over all measurement is allowed. Bumpers, limited to one-half inch overhang, are not included in the over all measurement.

Movable keels, metal fin keels or others without hollow garboard, centerboards, leeboards, bilgeboards, bowsprits and overhanging rudders are prohibited. There are no restrictions on scantlings or materials and no limit to displacement, *L.W.L.*, beam, draft, freeboard or tumble home.

No fore triangle measurement is taken but the actual sail area is measured. Reach of sails shall not exceed two inches. Rounded foot of loose-footed sails is not measured. There is no limit to the height of the mast but the height of the jibstay above the deck shall not be more than 80 per cent of the height of the headboard of the mainsail above the deck. The greatest diameter of a spar shall be ¾″.

Models of the Marblehead 50–800 Class are easier to build than the

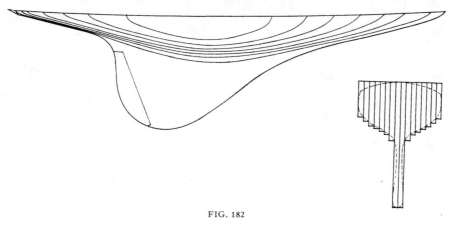

FIG. 182

larger ones of Class A but there are a number of "freaks" in existence in the class.

These racing models are not toys but sporting media worthy of any man's interest. In them one combines the three acts of designing, building and sailing the boat — usually impracticable in full size yachts. Much of value may be learned from the models.

Models six or eight feet over all may be built in various ways. The ambitious workman may want to plank up his model on frames, a method which is perfectly practicable. One way is to make a solid block model to the inside of skin, cut athwartships grooves in it to receive the steam bent frames, and plank on this rigid form which, of course, cannot spring. This is a big job but is particularly suited to making more than one model from the same design. Another method is to plank on molds erected upside down on the bench, or on web frames which are to remain in the hull. The deck should be fastened only at the sides to allow it to swell without buckling.

"Bread and butter," or lift models, are easier to make than planked models and are about as light. Models are usually made of horizontal lifts taken from the water lines, though vertical lifts taken on buttocks, as in Fig. 182, offer certain advantages. The longitudinal center line is automatically established; the two center lifts make a fine flat place to seize in the vise; the lifts are easily hollowed out before they are glued up. The two halves may be glued up separately and hollowed out to uniform thickness and smoothed on the inside before gluing together. Cross sectional templates should be used in bringing the outside to the finished surface. The sheer may be cut advantageously before gluing. Waterproof glue must, of course, be used.

INDEX